ASTROPHOTOGRAPHY
Second Edition – Revised and Enlarged

Featuring the fx system of Exposure Determination

ASTROPHOTOGRAPHY
Second Edition – Revised and Enlarged

Featuring the fx system of Exposure Determination

Barry Gordon

Willmann–Bell, Inc.
P. O. Box 3125
Richmond, Virginia 23235
United States of America

Publishers and Booksellers

Serving Astronomers Worldwide
Since 1973

Library of Congress Cataloging-in-Publication Data

Gordon, Barry, 1927–
  Astrophotography.

  Bibliography: p.
  Includes index.
  1. Astronomical photography. I. Title.
QB121.G67   1985   522'.63   85–12124
ISBN 0–943396–07–7

Printed in the United States of America
86 87 88 89 90 91 92 10 9 8 7 6 5 4 3 2 1

FOR JESSICA

PREFACE

This book grew out of a course offered at the American Museum—Hayden Planetarium in New York City. From time to time, students in this course would ask: Isn't there some book where we can find all this stuff? This book is intended to be the answer to that question.

The material collected for the course, and the book, came from a wide variety of sources over a period of several years. Some of it is drawn from the few existing books on the subject, and some from various works on astronomy. Much of the material has been gleaned from years of monthly magazines, and much from years of one-on-one discussions with knowledgeable, and helpful, workers in the field. Finally, there is material that is original and published in this book for the first time.

Because of the variety of sources and the time span involved, it is not possible to give credit to all whose contributions have found their way into this work—with the one mandatory exception of David Healy, to whom I am indebted for a great many things astrophotographic. Obviously, we all stand in debt to those who have gone before, as we hope that those who follow will benefit from our contributions.

WHAT THIS BOOK IS *NOT* ABOUT

One thing this book is not about is telescope-making. Or the construction of mountings. Or the making of any of the varied paraphernalia of the astrophotographer. These days, there is a plethora of commercially manufactured equipment of adequate quality. For the die-hard do-it-yourself-er, there is adequate literature detailing the how-to's of construction. This book assumes that all requisite gear is either now in hand, to be purchased, or to be built from instructions found elsewhere.

Another thing this book is not about is processing or darkroom technique, not that the subject is by any means trivial. There is much to be said on such topics as unsharp masking, integration printing, and other advanced processing techniques. This area, however, is simply outside the book's domain.

WHAT THIS BOOK *IS* ABOUT

This book is about capturing images of the heavens on film. Astrophotography has played a major role in advancing the science of astronomy. But to a great many of us, particularly the amateurs, the primary fact is that a good astrophotograph is a thing of surpassing beauty. This is the paramount consideration in our approach to the subject. We cannot escape the fact that, as astrophotographers, we are venturing into the realms of several sciences; astronomy, meteorology, chemistry, mechanics, optics, and mathematics. But our emphasis will always remain: How do we capture on film the spectacular beauty of the heavens?

When we must come to grips with mathematics — and we surely must if we are to do the planning good astrophotography requires — we shall attempt to make it as painless as possible, using a three-level approach. For the majority of our requirements, we shall use approximating expressions like $y = 3x$ or $z = x/y$. The second level — for those who demand complete precision — will be exact formulas of a slightly more complicated nature, but nothing beyond high school algebra and trigonometry. The final level — for those who enjoy that sort of thing — is derivations of the various formulas used; these are barred from the text and consigned to the appendices.

THE PHOTOGRAPHS

It would have been no trick at all to make this book a truly dazzling visual treat, using readily available reproductions of professional observatory photographs taken with the world's greatest telescopes. That temptation has been scrupulously resisted. In fact, even photographs produced in amateur observatories have been avoided, as have photographs produced using customized equipment fabricated by skilled craftsmen. Instead, what you will find here, exclusively, are photographs taken by amateur photographers/astronomers using standard, commercially available, portable equipment: in other words, photographs that could be taken by you.

CONTENTS

Chapter 1

INTRODUCTION

Astrophotography is the photography of astronomical subjects, that is, the entire universe beyond our earth's atmosphere, plus meteors. We shall consider our neighbors in the solar system: earth's moon, the sun, and the major planets. We shall also deal with our sun's peers, the other stars of our galaxy. Finally, we shall consider the so-called deep-sky objects: nebulae, star clusters, and the inconceivably remote galaxies beyond our own. While many of the astronomical terms will be defined as we come across them, the newcomer might do well to seek additional help in a good basic astronomy text.

CHARACTERISTICS OF ASTROPHOTOGRAPHY

The characteristic that most distinguishes astrophotography from all other photography is its range of variation. While far easier than you think it is, astrophotography is also more difficult than you can imagine. Spectacular examples have been produced by quick-witted photographers who have snatched their cameras and made "grab shots" with literally zero premeditation; equally spectacular, and rather more typical, examples were literally years in the planning and making. While the "terraphotographer" typically exposes film for 1/1000 to about one second, the astrophotographer uses all of the shutter speeds plus exposures of several seconds, or minutes, or even an hour or more. Compared to a brightness range of a mere 250,000 to 1, the astrophotographer tackles subjects with a brightness range on the order of 5,000,000,000,000 to one. To cope with these challenging subjects, the astrophotographer calls upon a range of optics in which the vaunted "super telephoto" of general photography ranks as no more than a lens of medium focal length.

Another characteristic of astrophotography, shared to a great extent by all photography, is the pervasiveness of trade-offs. Almost any problem we encounter can be solved, but time and again those solutions engender new problems.

1

A final characteristic to bear in mind is that this is a realm of exceptions. Again and again, we shall resort to hedges like: *most . . . in general
. . . as a rule*. When no such hedge is given, you are well advised to assume
that one has been omitted inadvertently.

REQUIREMENTS FOR (ASTRO)PHOTOGRAPHY

There are four requirements for *any* photography. The first of these is
light. Light is a scarce commodity in most astrophotography: but when
dealing with the sun, however, excessive light levels constitute a real and
serious danger.

Our second requirement is a subject. Unlike terraphotographers, we
deal with two very different kinds of subjects and two very different kinds
of subject brightness. Chapter 2 offers discussion of these.

Our third requirement is an optical system that consists of one or
more lenses, one or more mirrors, or a combination of both. As mentioned, astrophoto optics are frequently "long", but they run the gamut
from the so-called fish-eye to telephoto lengths inconceivable to the average photographer. We'll take up optics at much greater length in Chapters
2, 3, 4, and 8.

Photograph 1-1. A less-than-perfect attempt by an acknowledged
master—such things will happen. The subject is reported to be M45, the
Pleiades, taken October 1983 in Naco, AZ. Other data mercifully unre-
corded. By David Healy.

Our fourth and final requirement is a photographic emulsion. For most of us, this is just a fancy name for film, but the use of glass plates is common among professional astronomers. Film selection and handling raise many questions for the astrophotographer. For starters, we must select between color films and black-and-white, and, if we choose color, between negative and reversal (slide) films. We must weigh the trade-off of high "speed", or sensitivity, for our low light levels, against "fine grain" for maximum detail. These questions also will be discussed further in Chapter 2. Our last film decision, the choice between conventional and exotic handling, is discussed in Chapter 12.

Because available films and their specifications change from time to time, it is far more practical to stay current through the astronomy periodicals that serve the amateur than hope for specific data or recommendations from reference books.

The best bet for the beginner is probably a reversal color film, usually referred to as slide film, processed by the manufacturer or other good commercial laboratory, and customarily returned as mounted color slides. Since there is only one processing step between running the film through the camera and viewing the finished pictures, there is little opportunity to alter your work. With few exceptions, the slide is a reproduction of the exposure you made. On the other hand, printing astrophoto negatives is a challenge beyond the capabilities of most commercial labs. The non-do-it-yourself-er, then is far better off with color slides, which, in addition, may very well provide the absolute maximum return in sheer beauty.

A warning: *Do not send a roll of nothing but astrophotographs to any commercial laboratory, unless you specify that it not be cut.* Astrophotographs are foreign to photo-lab personnel, who just might slice up your film right through your exposed frames rather than between them. Subjects that are easily seen, like large images of the moon, are usually safe enough, particularly if sandwiched between conventional photographs that clearly indicate the locations of the frame edges. However, maximum safety lies in those three little words: *Do Not Cut.*

PREVIEW OF EQUIPMENT AND TECHNIQUES

One reason astrophotography is far easier than you may think is that a surprising amount of good astrophotography is possible without particularly elaborate, complicated, or expensive equipment. Many excellent astrophotographs, published in amateur astronomy publications, have been taken with common family-snapshot-type cameras using so-called normal lenses.

In Chapter 4, we shall discuss basic equipment: camera, lens, tripod — and notebook. Chapter 5 offers a (perhaps surprisingly) lengthy list of astrophoto subjects that can be photographed with this basic equipment, as well as how-to details. Subsequent chapters address more advanced equipment and techniques, including the equatorial mounting and guided exposures, and finally introduce some very advanced equipment and techniques for the astrophotographer of well above average dedication.

Advanced astrophotography is strictly Murphy's Law Country. It is a kingdom ruled by what David Healy calls "the fifty point handicap — the automatic reduction in your IQ that normally results from the attempt to go outside at night and take pictures through a telescope". This represents the harder-than-you-can-imagine end of the astrophotography spectrum, but that's a long way off. Here at the easier-than-you-think end, there is endless beauty to be preserved forever on film by the humble camera-lens-tripod combination.

Photograph 1-2. An early (October 1978) attempt with the Schmidt camera, illustrating many of the possible film-handling errors (film not flat in holder, film trimming stuck on film, film trimmed too small, one of the "stars" is a processing flaw, negative scratched, negative dusty, etc.). Orion's belt area on Kodak 103a·E, 300 mm $f/1.5$ Schmidt camera, 5 minutes. Taken in Naco, AZ by David Healy.

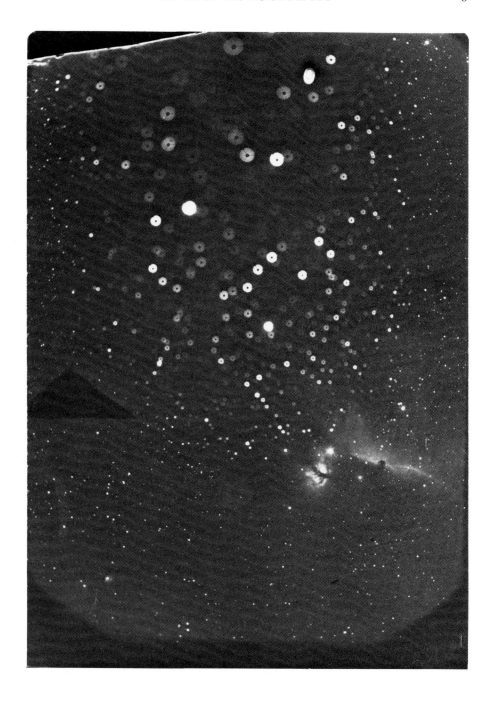

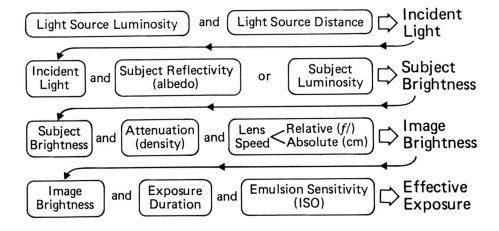

Figure 2-1. The Elements of Exposure.

Chapter 2
EXPOSURE

Let's face it: the subject of photographic exposure can be confusing. For the astrophotographer, it is just a bit more so, but with a modest amount of effort the subject can be mastered. This is where we make that effort toward achieving the mastery essential to all of our future endeavors.

THE ELEMENTS OF EXPOSURE

In photography, exposure means two different things: first, the act of exposing an emulsion to light and second, the image produced. The two meanings are closely related and it is rarely necessary to distinguish between them.

The three basic elements of exposure, i.e., the three factors that determine how much image-producing effect a given exposure will have are:

- Emulsion Sensitivity
- Duration
- Image Brightness.

Image Brightness, in turn, is a result of three independent factors:

- Subject Brightness
- Attenuation
- Lens Speed.

These are shown schematically in Figure 2-1.

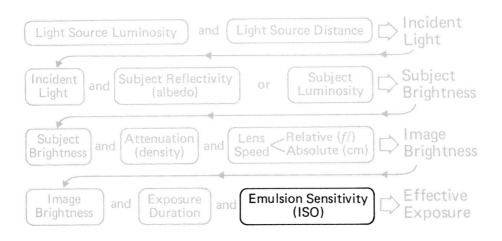

Emulsion Sensitivity

Emulsion sensitivity is simply a measure of an emulsion's sensitivity to light. As shown in the diagram, it is the very last factor — added to the actual exposure itself — that determines the effect of that exposure. Because most of us use film instead of glass plates and because a more sensitive film forms a latent image faster than a less sensitive one, the common term for emulsion sensitivity is film speed.

Film Speed

For many years, the standard method in the United States for specifying film speed was the ASA rating, named for its creator the American Standards Association (now the American National Standards Institute). ASA ratings for typical films ran in the 25 to 400 range, and were proportional to the film speed, e.g., ASA 400 film was twice as fast as ASA 200 film which was twice as fast as ASA 100 film and so on. There were also funny in-between numbers that didn't fit the 2:1 ratio scheme; these simply represented in-between film speeds, which were still proportional to their ASA ratings.

Independently, the Germans established a "German Industry Standard" system — Deutsche Industrie Norm, or DIN for short — which you might still run across from time to time. They used a logarithmic film speed scale in which, for example, DIN 15 was equivalent to our ASA 25, and each increase of three represented a doubling of speed. Therefore: DIN 18 = ASA 50, DIN 21 = ASA 100, and so on.

The next development in film rating was an ISO (International Standards Organization) system. Originally a combination of both the

ASA and DIN ratings, its designations appeared in that order, separated by a slash. That system was short-lived (thank heaven), and now film speeds are designated by ISO ratings, which are identical to our earlier ASA ratings. A former ASA 64 film is now an ISO 64 film, and so on.

Because ISO 200 film photographs in one second what would take ISO 25 film eight seconds, one might legitimately wonder why we bother at all with any but the fastest films. The answer is resolution. Put simply, slower films tend to have finer emulsion granularity, or finer grain, and thus render finer detail than faster, coarser-grained films. If we are able to make our exposure in eight seconds, we will probably get a better quality image with ISO 25 film; if for any reason, our exposure limit is one second, then the ISO 200 image certainly beats losing the shot.

Reciprocity and Reciprocity Failure

The above discussion introduces another vital point: reciprocity. This simply refers to turning a number upside down. Thus, from the example above, cutting film speed to one eighth calls for eight times the exposure duration. This so-called law of reciprocity holds for all of the various exposure factors, i.e., halving one factor is compensated for by doubling another. Generally!

Reciprocity failure, a pitfall almost unknown to the average photographer, lies in wait to trap the unwary astrophotographer. Doubling exposure time will counterbalance halving the light — up to a point. Beyond that point, the increase in one will not quite make up for the decrease in the other. Creating a latent image on film is somewhat analogous to opening a door; it requires a meaningful push for a meaningful period of time. You achieve nothing by an instantaneous sledge-hammer blow, and just as much by blowing on it for an hour. So it is with film. Most film is intended for fairly decent light levels and exposures in the 1/1000 to 1/10 of a second range — the kind of use accommodated by the ISO ratings. But exposed to starlight for half an hour, a typical ISO 400 film may act more like ISO 50 or worse. Another way of looking at it is that, beyond a certain point, doubling exposure duration will do much less than double the effect of the exposure, and doubling it again will do still less.

There are three approaches to reciprocity failure. The first is simply to understand it and live with it. The second is to use a special astrophoto film designed to retain its sensitivity when exposed for long durations at very low light levels. These black-and-white negative films are available, but not through normal photo outlets. (These films have no official ISO ratings, though we shall make assumptions as necessary in order to provide quantitative exposure data.) The third approach to the problem involves rather advanced equipment and/or techniques and will be taken up in Chapter 12. For most of our discussion we will settle for the first approach.

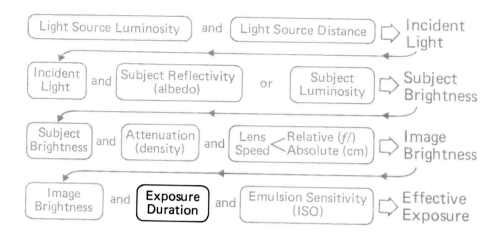

Duration

Continuing to work backward in our schematic, the next simple element of exposure is duration. Exposure duration is the length of time the film is exposed. To the terraphotographer this is usually synonymous with shutter speed, commonly 1/1000, 1/500, . . . , 1/2, 1 second. As stated earlier, the astrophotographer uses all of these "built-in" shutter speeds, plus durations of many seconds or minutes or possibly an hour or more (very often not using the shutter at all). One implication of this is that a good solid tripod is essential to the astrophotographer, something many photographers do without. A rule of thumb instructs not to hand-hold a camera for durations, in seconds, longer than the reciprocal of the lens focal length in millimeters. In other words, use 1/60 or faster with a normal 50 mm lens, 1/125 or faster with a 135 mm telephoto, 1/500 or faster with a 500 mm telephoto. When we start looking at the lens lengths and exposure durations we'll be wanting to use, the need for a tripod will be obvious.

Unfortunately, that's only the beginning. We all know that the tripod's function is to keep the camera "motionless" during the exposure. But the very best tripod can do no better than keep the camera motionless with respect to the earth. To keep our camera motionless with respect to our subject, we must move it in a very precise way with respect to the earth. This, however, gets us beyond basic equipment and into Chapter 6. The point to be made here is simply that it is exposure duration that determines the camera mounting we require—our hands or a conventional tripod or something more elaborate; and we will pin that down more precisely in Chapter 3.

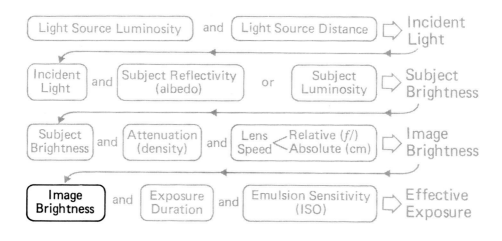

Image Brightness

Image brightness is the brightness of the image that the lens forms upon the film surface. This, however, is not a simple thing. As stated earlier and indicated on our diagram, it is the result of subject brightness, attenuation, and lens speed—each of which bears closer scrutiny.

Subject Brightness

A subject is brighter or dimmer because of its own light, or because of the light it reflects. A subject of the first kind is called luminous (or, redundantly, self-luminous). Luminous subjects include bulbs and bonfires, street-lights and stars. The brightness of a luminous subject depends on nothing more than its luminosity, i.e., how much light it generates. A subject of the second kind is called nonluminous. Nonluminous subjects include pebbles, people, and planets; they are the subjects of the upper left portion of our exposure schematic. The brightness of a nonluminous subject depends first on how much light it receives, called incident light, from its source of illumination (which, in turn, depends on that source's brightness and distance), and then on the proportion of that incident light that it reflects, called its albedo. Here on earth—and, for that matter, throughout the solar system—all of us nonluminous subjects are illuminated by the sun (ignoring for the moment, artificial light). Coal piles look darker than snow banks, because coal reflects less light. Albedo, then, is simply the amount of received light that a subject reflects, theoretically ranging from zero to 100%. Coal runs well below 10%, and clean snow well over 90%; an average earth landscape has an albedo of 18%, and photographic exposure meters are calibrated on that basis. Planets are somewhat more complicated, having different distances from their light source in addition to their different albedos. However, in practice, this is of no concern, since we generally deal with subject brightness as a single simple factor, irrespective of how it actually comes about.

There is another dual aspect of subject brightness that we cannot ignore. Brightness is two different things. If someone shines an ordinary flashlight in your eyes, the effect is dazzling, perhaps even mildly painful. However one experiences no discomfort when looking at a common multi-tube fluorescent fixture, such as might be found in one's kitchen. Obviously, then, the flashlight is brighter. But, you cannot light your kitchen with just a flashlight. Thus, the fluorescent fixture puts out more light. Is this a contradiction? Not at all. No more than a match burning your hand while the radiator can heat a room: higher temperature on the one hand (the burned one), more heat on the other. In similar fashion, the astronomer says that the full moon is "brighter" than Venus, while the photographer knows that Venus requires less exposure than the full moon. The answer is that the full moon, appearing thousands of times the size of Venus, gives off much more total light, but any portion of the full moon having the same apparent size as Venus would appear much dimmer. The astronomer's "brightness" generally refers to total light — that is what he means by magnitude. On occasion, to help avoid confusion, he calls it integrated magnitude. The photographer's "brightness" is referred to in astronomy as surface brightness. Thus, the full moon has the greater magnitude, but Venus has the greater surface brightness. This important distinction applies only to what we call "extended objects" (defined later in this chapter).

Attenuation

Having discussed subject brightness and image brightness, we can now examine what comes between: obviously the lens, possibly something else. Often, this is a filter, typically a piece of colored optical glass placed before the lens to enhance contrast, for example. However, filters "steal" some of the light on its way from subject to image, and this should be taken into account. Since they do attenuate image brightness, they are called attenuators. As mentioned, and as shown near the center of our diagram, attenuation is the second of the three elements that determine image brightness. True filters attenuate selectively, blocking more of one kind of light than another, depending for example, on color or direction of polarization. Neutral density attenuators (usually but incorrectly called neutral density filters) do not filter or act selectively, but rather attenuate all light uniformly. Their main use is in photographing the sun. While

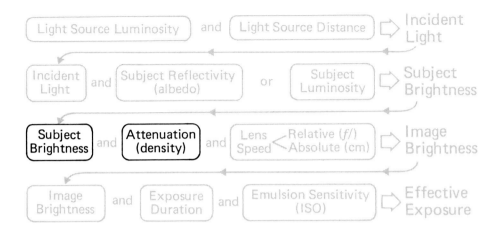

weaker filters and attenuators are often described by "filter factors", they are generally rated by density. Each density increase of .3 cuts transmission by half and thus calls for a doubling of the exposure to compensate. A density of zero indicates total transmission, i.e., no attenuation at all. For example, an attenuator with a density 1.5 (which is five times 0.3) would call for 32 times (i.e., five doublings of) the exposure otherwise required.

A final word on attenuators (until we deal with solar photography): there is one attenuator that astrophotographers deal with all the time, like it or not. It is our atmosphere. We don't think about it very much, because most of our exposure data already include its effect to some extent. We should, however, be aware that the effects of atmospheric attenuation are significantly aggravated as we move lower in the sky. As a result, near the horizon, we "lose" stars that are clearly visible higher in the sky. The astronomer's term for this effect is atmospheric extinction. When dealing with brighter subjects, we can generally compensate for it, but photography of dimmer subjects is best done when they are well above the horizon.

Lens Speed

The last of the three image brightness elements is lens speed. As stated in Chapter 1, our optics can consist of lenses or mirrors or combinations of the two, though for simplicity, we will refer to them all by the general term lens. As far as speed is concerned, the lens has only two significant attributes: aperture and focal length. The aperture is the lens opening or "window" through which light is admitted. Very often, but not always, the aperture is the front surface of the lens, a lens measuring 10

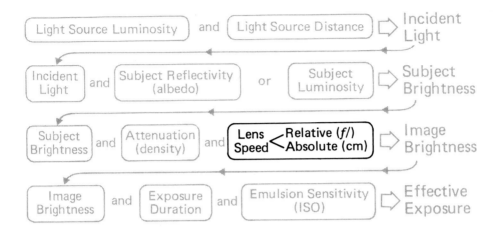

cm across its front surface having an aperture of 10 cm. In some cases, wide-angle lenses, for example, aperture and front surface can be very different. For the lenses we astrophotographers use most of the time, the two are the same.

In the early days of photography, when photographers wanted to decrease their lens apertures, they used an aperture stop — a sheet of metal with a suitable hole in it, stopping all the light except that which went through the hole. In time, however illogically, the word "stop" came to apply to the hole itself. Today, the word "stop" is still used to refer to the aperture, which we generally vary by means of an iris diaphragm.

Lens focal length is the distance from the lens to its image plane, i.e., the plane in which the image comes to its focus. (Strictly speaking, the definition should specify that we mean the image plane for an "infinitely distant" subject; but, as astrophotographers, that's the only kind of subject we consider.) Both aperture and focal length are directly measurable. They can be expressed in various units of length; we will tend to use millimeters most of the time, and centimeters occasionally.

From these two basic measurable attributes, we can calculate a third attribute, one that has no physical existence. Though it cannot be seen, much less directly measured, it is extremely useful. This is the focal ratio, and it is simply the focal length divided by the aperture. Using their obvious letter symbols:

$$r = \frac{f}{a}$$

To calculate the focal ratio, r, we must specify f and a in the same units of measure; the ratio r itself is simply a number, in no units at all.

Now let's see what these things do. First and simplest, the lens aperture, a, lets in the light. How much light does it let in? How much water goes through a pipe? Other things equal, the bigger the pipe diameter, the more water. And so it is with a lens: other things equal, the bigger the aperture, the more light. In fact, the amount of light available increases with the square of the aperture — double the aperture and you quadruple the available light, triple the aperture and you get nine times the light. In the (usually) light-starved world of the astrophotographer, aperture is of prime importance.

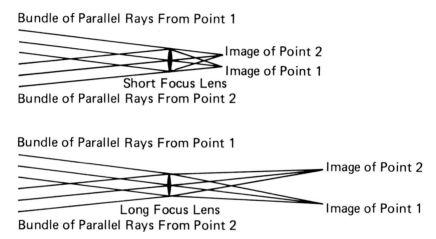

Figure 2-2. Image Size Variation with Focal Length

What about focal length? We said that focal length is the distance to the point where the subject comes to focus. That means that the light rays from star #1 (see Figure 2-2), after covering the entire lens, now converge at distance f behind the lens. The same thing is happening with the bundle of rays from star #2. This is true whatever the value of f. The two points toward which those rays are converging will, of course, be two separate points in the focal plane, at some distance from each other. The distance between those two points will be proportional to the focal length of the lens. Further, the same process occurs for light from two different points on the same subject, as well as two different stars. The further those rays travel from the lens before converging, the farther apart will be their points of convergence. In other words, image size is directly proportional to f, the focal length of the lens. This is illustrated by the diagrams in Figure 2-2 and by Photographs 2-1.

Photographs 2-1. The moon, about two days before full. Taken May 5, 1979, in New York City, on Kodachrome 64. In order, exposures were: 50 mm f/8, 1/60 sec; 135 mm f/8, 1/60 sec; 300 mm f/8, 1/60 sec; 750 mm f/6, 1/60 sec; and 1500 mm f/12, 1/15 sec. Total $fx = 12$.

That's just great: if we want an image twice as big, we simply use a lens with twice the focal length. That, of course, is precisely what telephoto lenses are all about. In fact, it is of such basic importance that focal length is the first, and often the only, lens attribute that a photographer gives. Thus, the normal 50 mm lens is a lens of 50 mm focal length, the 200 mm telephoto has a 200 mm focal length, and so on. (The light-obsessed astronomer, on the other hand, talks exclusively of aperture; what he calls a 200 mm lens would have a 200 mm aperture. This is one more point of confusion between astronomer and photographer, with the astrophotographer, as usual, caught in the middle.)

Before we start thinking we're getting something for nothing, let's look into the cost of these lovely large images. Keep in mind that we're discussing the subject's image formed by the lens, totally independent of the film. How much light does such an image require? How much paint does a wall require? Other things equal, the bigger the wall, the more paint. So it is with an image: other things equal, the bigger the image the more light is required. In fact, the amount of light required increases with the square of the image size — double the image size and you quadruple the required light. By now, this wording should look very familiar.

Now where is all of this getting us? Actually, it's getting us right where we want to go: lens speed as an element of image brightness. If we want lens speed to mean the ability to make exposures rapidly, since brighter images tend to require shorter exposures, then a "faster" lens should be one that produces a brighter image. A brighter image results from increasing aperture, or from decreasing focal length (and thus decreasing image size). Further, brightness of image varies according to the square of each, but in the opposite direction. Consequently, if we double, or halve, or do anything the same to both aperture and focal length, the image brightness remains the same. In short, the one thing that determines lens speed is the focal ratio, r. All lenses with a focal ratio of 8, for example, will give the same image brightness in any given situation; actual aperture and focal length don't matter, as long as the latter is eight times the former.

Thus, instead of giving the actual aperture of a lens in centimeters or the like, the photographer finds it much more useful to specify it in terms of focal ratio. Note that, since r = f/a (by definition of r, as noted earlier), it is obviously also true that a = f/r. Thus, continuing with our example of r = 8, we would say that our lens has an aperture of $f/8$. As stated above, all lenses with a = $f/8$ are equally fast. Apertures of $f/4$ are four times this fast, and apertures of $f/2$ are four times as fast as that. We refer to apertures specified this way as relative apertures, as distinguished from absolute apertures given in, say, centimeters, for example. With the relative aperture, or its corresponding focal ratio, firmly established as the criterion of lens speed, we arrive at the now familiar standard scale

$$f/1 \quad f/1.4 \quad f/2 \quad f/2.8 \quad f/4 \quad f/5.6 \quad f/8 \quad f/11 \quad f/16 \ldots$$

and to understand how the term "stop" evolved into "f/stop".

EXPOSURE DETERMINATION

At this point, even though there is one item in our exposure schematic that we've yet to discuss, you know almost all you need to know — and a great deal more than most photographers — about what goes into the photographic exposure process. Now let's see how we can put it all together to obtain properly exposed photographs. In summary, the elements of exposure (with symbols to make discussion simpler) are:

Emulsion Sensitivity	I for ISO
Duration	t for time, generally in seconds
Image Brightness	
Subject Brightness	B
Attenuation	D for density
Lens Speed	r for focal ratio

Since image brightness is merely the net result of the last three items, we have not given it a symbol. Of the five we must deal with, I, t, D, and r are pretty much under our control: B is the given, to which we must adjust the others. We will discuss two different, but related, ways of doing this. But first, we must dispose of a common myth.

The myth goes by the name of "Correct Exposure". Even if there were such a thing, no exposure system or device could possibly assure obtaining it. For one thing, ISO ratings are not that exact; a nominal ISO 100 film, for example, could actually be closer to ISO 80 or ISO 125, varying from batch to batch. Camera shutter tolerances also enter the picture, with variations of up to 50% not uncommon. And it's just not true that all $f/8$ lenses have exactly the same speed, what with different types of construction giving different amounts of internal reflection and absorption. On the whole, these variations tend to balance out, and we go on our way blissfully ignorant of the difficulty of our task, and getting perfectly good results most of the time. Slight variations are almost always present, and every now and then they gang up in the same direction and throw things off the mark.

Even if our chemistry and mechanisms and optics were absolutely perfect, "Correct Exposure" just does not exist. To your light meter, the world has an albedo of 18% and therefore looks medium gray — coal piles to snow banks — and that's the way it will try to make everything look in your photographs. So light meter correct is not necessarily your idea of correct, and your idea of correct may not be someone else's idea of correct. You can take pictures, for example, of landscapes by moonlight. "Correct" exposure will reveal details as plain as day, but the problem is: by moonlight things just don't look as plain as day. So who is correct? The light meter? You? Someone else? Answer: all of the above. What does the moon look like? Is it bright silver-white, or is it dull gray-brown? Yes, it is: the former in distant views showing the entire moon against a dark sky; the latter in ultra-close-ups (e.g., those taken on the spot) showing only a small piece of the landscape. Same moon. Very different exposures. Both correct. In astrophotography the myth of correct exposure is especially out of place. Many subjects have whole ranges of correct exposures, with lesser exposures showing inner detail and greater exposures showing outlying extent, simply because many subjects have a range of brightness. In the final analysis, the correct exposure is simply the exposure you want, and no system or mechanism can tell you what that is. What we can do, however, is set down exposure recommendations, i.e., starting points, based on experience. These recommendations will serve quite well to get you "off the ground," and in many cases will probably suffice. However, they make no pretense at correctness.

Our first attack on the exposure problem is an algebraic one that goes back many years. We call it the BItrD approach, and it ties our five elements together as follows:

$$B \; I \; t \; = \; r^2 \; 10^D$$

That's the whole thing. But since we very often do without filters or other attenuators, and since $D = 0$ makes 10^D equal to 1, we can simply throw away the entire last factor. Note that this factor becomes vitally important for solar photography, as we will see, but for other purposes we can usually get by with:

$$B \; I \; t \; = \; r^2$$

This will handle most of our problems. For example, suppose we know — ignoring for the moment how we know — that our subject has a brightness value, B, of 250, and we wish to photograph it on ISO 64 film using our $f/8$ telephoto (remembering that $f/8$ is the relative aperture, the focal ratio is 8). Substituting our known values:

$$250 \times 64 \times t = 8^2$$

which tells us that we should expose for 1/250 of a second. That's all there is to it. Following this formula what ISO film do I need in order to photograph a B = 0.063 subject at $f/8$ (ratio = 8), keeping my duration to 1 second?

$$0.063 \times I \times 1 = 8^2$$

Ignoring the rounding imprecision, it seems clear that we need ISO 1000 film for this job.

Even in this age of hand-held electronic marvels, many people are still happiest with the least possible mathematics. So alternatives to the BItrD approach to exposure determination were inevitable. One approach is the use of exposure tables. However, even leaving out attenuation, we still have four variables to cope with, a challenge to simple presentation in tabular form. Occasionally, a sort of "double table" can be used to good effect, and we will certainly take advantage of those occasions as they arise. However, a good, simple, generally applicable technique should be possible.

The approach we shall adopt was developed by the author specifically to facilitate the discussion and determination of exposure for the astrophotographer. Obviously, since terraphotography deals in a subset of the exposure problems of astrophotography, a subset of this exposure system will handle terraphoto exposure problems quite readily. The system is called the *effective exposure* system, or *fx* system for short.

THE *fx* SYSTEM

The basis of the *fx* system is simply this: Every photographic subject calls for a total effective exposure, and each of the photographer's variables (film speed, duration, attenuation, lens speed) makes its own effective exposure contribution to that required total. (The term "effective exposure" is used because while duration, attenuation, and lens speed determine the actual exposure, film speed merely determines the effect of that exposure.) The scale of values adopted for the system was chosen to satisfy the following requirements:

1. A larger *fx* value indicates a greater effective exposure
2. A difference of 1 in *fx* value indicates a difference equivalent to 1 *f*/stop in effective exposure, i.e., a halving or doubling of the effective exposure
3. All exposure problems are solved in the system by addition and/or subtraction of simple numbers

    Table 2-1 shows a basic table of *fx* values for the four variables under the photographer's control: the columns of the table correspond to the four basic elements of exposure. Since any given subject will call for just one total *fx* value, any duration-aperture-film-attenuator combination, whose individual *fx* values sum to the total *fx* required by the subject, represents the recommended exposure for that subject.

fx	Second(s)	*f/*			ISO			D	*fx*
17	30								
16	15								
15	8								
14	4								
13	2								
12	1								
11	1/2								
10	1/4	1.1	1						10
9	1/8	1.6	1.4	1.3					9
8	1/15	2.2	2	1.8	3200	4000			8
7	1/30	3.2	2.8	2.5	1600	2000	2500		7
6	1/60	4.5	4	3.6	800	1000	1300		6
5	1/125	6.4	5.6	5	400	500	650		5
4	1/250	9	8	7	200	250	320		4
3	1/500	13	11	10	100	125	160		3
2	1/1000	18	16	14	50	64	80		2
1	1/2000	25	22	20	25	32	40		1
0	1/4000	36	32	28		16	20	0.0	0
-1		50	45	40				0.3	-1
-2		70	64	56				0.6	-2
-3		90	80					0.9	-3
								1.2	-4
								1.5	-5
								1.8	-6
								2.1	-7

Table 2-1. A Basic Table of *fx* Values. For most purposes, multiple entries in a row may be treated as having the same *fx* value. When greater precision is desired, note that entries left or right of the central one have *fx* values 1/3 lower or higher, respectively.

Now let's go back and redo those two earlier exposure problems. What is the duration for a subject with $fx = 10$ — again ignoring how we know this — on ISO 64 at $f/8$? The fx contributions (from the table) for our film and aperture are 2 and 4, respectively. To reach the required 10, we need 4 more, which we get from a duration of 1/250 of a second. Our second subject has $fx = 22$. An exposure of 1 second at $f/8$ gives fx contributions of 12 and 4, respectively. The 6 we lack will be contributed by a film rated ISO 1000. Note that the fx contribution for a $D = 0$ attenuator, i.e., no attenuator at all, is itself zero and thus drops out very nicely. Thus, in most cases, exposure problems are solved by finding three numbers to sum to a fourth, all of them simple numbers within a very limited range.

Well, that's all very well and good, you say, but just where do we get these B values and/or fx totals needed to solve our exposure problems? Answer: right here. As we discuss the various subjects for our astro-cameras, we will be providing fx totals for all likely subject types. For those with an algebraic bent, Appendix A provides the formula for converting these into B values. Remember we offer no guarantees of correct exposure; just good values, based on experience, to get you started. As a bonus, Table 2-2, based on data from Kodak* should solve many of your earth-bound exposure problems.

We've said that image size is directly proportional to focal length. For the terraphotographer, that's fine; for the astrophotographer, we have to add the qualifier "as a rule," since we deal with literally countless subjects that violate that rule — stars. Most stars are so distant, with resulting apparent sizes so small, that no amount of focal length will create any enlargement in their images. Such subjects are called point sources to distinguish them from the so-called extended objects (which do obey the rule that image size is proportional to focal length). In fact, the simplest way to define a point source might well be to say it is any subject whose image size is not directly proportional to focal length. Note that we have not said that point source images are themselves points; we have not even said that they are all the same size. The fact is, as casual examination of an astrophotograph will reveal, different stars produce different sized images. The image size produced depends on a number of factors, notably the star's magnitude. Another factor is exposure duration. The key fact is that image size is not directly proportional to focal length. That is the essence of a point source. What it means to us is that point source exposures depend on absolute, not relative, aperture. In other words, lens speed means relative aperture for extended objects, but absolute aperture for point sources. That, of course, explains the presence of the hitherto ignored item, absolute aperture, in our schematic of exposure elements. In either case, greater aperture — relative or absolute — means greater speed.

---

*"Adventures in Existing-Light Photography", Kodak Publication AC-44, May 1978; plus "Photonews", Kodak Periodical AI-2-83-3, 12-83; plus film boxes.

Time/Place	Subject(s)	_fx_
Outdoors	Sunlit Snow or Beach Scenes	9
by Day	Average Sunlit Subjects	10
	Cloudy Bright Scenes — No Shadows	12
	Heavy Overcast or Open Shade	13
Home	Areas with Bright Light	18
at Night	Areas with Average Light	19
	Christmas Lighting or Trees	20
	Close-ups by Candlelight	20
Outdoors	Sunrises, Sunsets	12
at Night	Neon and Other Lighted Signs	15
	Store Windows	16
	Bright Nightclub and Theater Districts	16
	Brightly Lighted Downtown Scenes	17
	Fairs and Amusement Parks	18
	Subjects Lighted by Streetlights	20
	Skyline — Ten Minutes After Sunset	15
	Floodlighted Buildings, Fountains, etc.	20
	Distant Skyline, Lighted Buildings	23
	Burning Buildings, Bonfires, Campfires	16
	Subjects by Bonfires, Campfires	19
	Fireworks Displays on the Ground	16
	Aerial Fireworks — Snapshots	17
	Aerial Fireworks — Time Exposures (see note)	6
	Lightning (see note)	7
	Niagara Falls — White Lights	23
	Niagara Falls — Light Colored Lights	24
	Niagara Falls — Dark Colored Lights	25
	Moonlighted Landscape	27
	Moonlighted Snowscape	26
Indoors	Night Football, Baseball, Races	16
and/or	Basketball, Hockey, Bowling	17
Public	Boxing, Wrestling	16
Places	Stage Shows	16
	Circuses — Floodlighted Acts	17
	Circuses — Spotlighted Acts	15
	Ice Shows — Floodlighted Acts	16
	Ice Shows — Spotlighted Acts	15
	Interiors with Bright Fluorescents	16
	School Stages and Auditoriums	19
	Swimming Pools — Above Water	18
	Churches — Artificial Light	19

Table 2-2. _fx_ Totals from Data by Kodak. Note: For time exposures of aerial fireworks and lightning, _fx_ total is from _f/_ and ISO only.

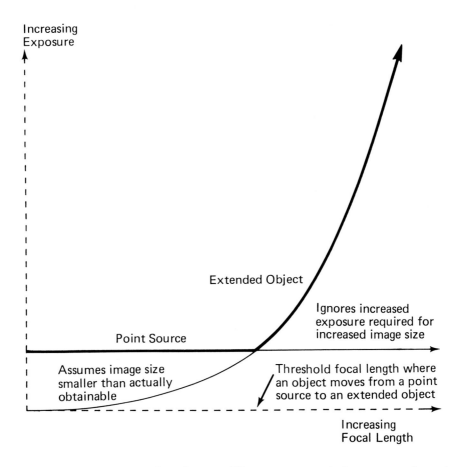

Figure 2-3. Exposure for planets. The recommended exposure is to be found along the portion of the curve which is plotted with a darker line.

So now we know our subjects form two kinds of images that must be treated differently. We know that most stars act like point sources, and we know that most other things — including the sky itself — act like extended objects. That would wrap it up fairly well, except for the "switch-hitters": our solar system neighbors, the major planets, which insist on having it both ways. Recall that a point source does not produce a true point image. As noted, image size depends on various things, including lens characteristics, film grain size, and the very nature of light itself. For each combination of the various factors, there is a theoretical minimum image size for a point source; in practice, the achievable minimum image is significantly larger than that. In other words, starting with subjects of the right (or wrong, depending on your viewpoint) size, decreasing focal length will decrease image size in proportion, down to a critical threshold image size, which will then remain essentially constant despite any fur-

ther focal length decrease. A good practical working value for this threshold image size is 0.1 mm. It just happens that the major planets are all
about the right apparent sizes to cross this image size threshold within the
range of focal lengths commonly employed by the astrophotographer.
Thus, when we photograph the planets, before determining exposure, we
must first determine, for the focal length we plan to use, whether the
particular planet is a point source or an extended object, as illustrated in
Figure 2-3. We'll pin this down properly in Chapter 9.

Note that the BItrD approach is for extended objects only. That is
why the formula involves focal ratio r and totally ignores absolute aperture. Further, our brightness values B refer to surface brightness. This is
an attribute of extended objects only; point sources are generally referred
to by magnitude, which is also lacking in our formula. While the algebraic approach could be used (with a separate formula, given in Appendix
D) for point sources, we will not go that route. Rather, we will rely on the
fx approach, which is easier to use and which deals with all subjects in a
more uniform manner. Table 2-3 shows the complete *fx* values which will
be the basis for our exposure data. This system will prove handy in a
surprising number of ways, the simplest being the solution of exposure
problems, as we have seen. For those readers who want all the details,
Appendix A gives the mathematical foundations on which the system is
based; these are totally unnecessary for the system's primary use, but
required for understanding certain derivations given in other appendices.

fx	Duration				f/			cm			ISO			D	fx
25	100	120		minutes											
24	50	60	80												
23	25	30	40												
22	13	15	20												
21	6	8	10												
20	3	4	5												
19	1.6	2	2.5												
18	50	60	80	seconds											
17	25	30	40												
16	13	15	20												
15	6	8	10												
14	3	4	5												
13	2					Aperture									
12	1					f/		cm							fx
11	1/2										ISO				
10	1/4			1.1	1		900	1024							10
9	1/8			1.6	1.4	1.3	640	700	800	6500	8000				9
8	1/15			2.2	2	1.8	450	512	560	3200	4000	5000			8
7	1/30			3.2	2.8	2.5	320	360	400	1600	2000	2500			7
6	1/60			4.5	4	3.6	220	256	280	800	1000	1300			6
5	1/125			6.4	5.6	5	160	180	200	400	500	650			5
4	1/250			9	8	7	110	128	140	200	250	320			4
3	1/500			13	11	10	80	90	100	100	125	160			3
2	1/1000			18	16	14	56	64	70	50	64	80		D	2
1	1/2000			25	22	20	40	45	50	25	32	40			1
0	1/4000			36	32	28	28	32	36		16	20	0.0		0
-1				50	45	40	20	22	25				0.3		-1
-2				70	64	56	14	16	18				0.6		-2
-3				100	90	80	10	11	13				1 0.9		-3
-4				140	128	110	7	8	9				1.2		-4
-5				200	180	160	5	5.6	6.4				1.5		-5
-6					256	220	3.6	4	4.5				1.8		-6
-7							2.5	2.8	3.2				2.1 2		-7
-8							1.8	2	2.2				2.4		-8
-9							1.3	1.4	1.6				2.7		-9
10								1	1.1				3		-10
													:		:
													4 3.9		-13
													:		:
													5.1 5		-17
													:		:
													6		-20

Table 2-3. Comprehensive Table of *fx* Values. For most purposes, multiple entries in a row may be treated as having the same *fx* value. When greater precision is desired, note that entries left or right of the central one have *fx* values 1/3 lower or higher, respectively.

CHAPTER 2 EXERCISES

The answers to these exercises are given, and explained, at the back of the book. Note: Quantities in mm are focal lengths.

1. Your subject, perhaps the gibbous moon, has fx = 12. What is your exposure duration with ISO 64 film and a 135 mm $f/2.8$ lens?

2. Your subject, perhaps Venus — as an extended object, has fx = 8. What is your exposure duration with ISO 64 and a 750 mm $f/8$ lens?

3. Your subject, perhaps Jupiter — as an extended object, has fx = 13. What is your exposure duration with ISO 64 and 750 mm $f/8$ lens?

4. With a 750 mm $f/8$ lens on a standard (but sturdy!) tripod, you want to keep exposures to 1 second. What film do you need to shoot a subject, possibly a lunar eclipse, with fx = 23?

5. Your subject is stars — repeat: stars — of magnitude 3. Assuming fx = 13, what is your exposure duration on ISO 200 film with a 135 mm $f/2.8$ lens?

6. Again, your subject is stars of magnitude 3. Assuming fx = 13, and a 1/2 second exposure on ISO 1,000, what must be the focal ratio of your 1500 mm lens?

Chapter 3
IMAGE SIZE AND RELATED DATA

As we saw in Chapter 2, in general (that is, for everything except point sources), image size is directly proportional to lens focal length: doubling, or tripling the focal length will double or triple the size of the image. We will now see how to determine image size and related items in advance, so we can select the appropriate focal length. But first, let's review our measuring scheme.

The basic unit of measure for the astrophotographer is the degree, for which we use the ° symbol. A degree is simply 1/360 of a circle—any circle. This could be a source of possible confusion. On earth, the system of longitudes and latitudes is also based on degrees. One degree of latitude is always about 111 km; which is 1/360 of the earth's roughly 40,000 km circumference. A degree of longitude, however, is only that big at the equator; it decreases as the circles it is based on decrease. On earth, this poses no problem; we measure distances in kilometers and the like, which obligingly stay one size.

Astrophotographic measurements are not quite so easy. In the first place, we don't care how big or how far apart things really are. Knowing the diameter of the moon in kilometers is useless to us, as is knowing the actual distance between two stars, one of which may be many times farther away than the other, though they appear quite close together. Worse yet, what could possibly be meant by the "size" of a constellation?

We deal with our measurement problem by first inventing an imaginary sphere of infinite radius, with us at its center. We call this "the celestial sphere" and pretend that everything in the sky is somehow affixed to it. On this sphere, the degree is our standard measuring unit, but we must remember that it comes in various sizes. The degree we will use most often is the largest of all possible degrees. It is, as are all degrees, 1/360 of a circle. But these full-size degrees are based on what we call great circles, so called because they are the largest circles possible in their spheres. Celestial great circles can also be identified by the fact that we are at their common center.

Getting back to earth, perhaps the most famous of all great circles is earth's equator. Other well-known great circles are the meridians of longi-

tude. The projections of these great circles onto the celestial sphere—
called the celestial equator and the meridians of right ascension,
respectively—are also great circles. Countless others exist, but have no
names. The full-size degrees of such great circles are what we mean when
we speak of our subjects' apparent sizes.

When we need smaller units than the degree, we will use 1/60 of a
degree. We call this a minute, and use ′ for its symbol. When this is still
too large, we will go to 1/60 again, or 1/3600 of a degree; this is called a
second and ″ is its symbol. Thus: $60″ = 1′$ and $3600″ = 60′ = 1°$.
Occasionally, when necessary to avoid confusion with time units, we refer
to arc minutes and arc seconds, also written arc min and arc sec. (As
astrophotographers, such confusion should only be our biggest problem.)

To give an idea of what some of these angular sizes mean, Table 3-1
gives a sky sampler. For those not conversant with star lore: Polaris is the
pole star, within one degree of the north celestial pole. Merak and Dubhe
are the two "pointers" to Polaris, with Dubhe, the nearer, 29° away.
Alkaid marks the end of the Big Dipper's handle, so the Alkaid-Dubhe
distance is the Dipper's size. Mizar is the middle star in the Big Dipper's
handle, and Alcor is its neighbor, the two often referred to as horse and
rider. Deneb and Albireo are, respectively, the "top" and "bottom" of the
Northern Cross, so this distance is the "height" of the Northern Cross.
Betelgeuse and Rigel are Orion's brighter shoulder and knee, so this
distance is the size of that figure's major portion. Alnitak and Mintaka are
the ends of Orion's belt. Appendix B derives the relevant formula.

The objects with the peculiar names, e.g., M42, are "deep-sky"
objects designated by their Messier Catalog numbers (discussed further in
Chapter 11).

Jupiter and Neptune illustrate the relative sizes of (apparently) large
and small planets. They differ by a ratio of about 20 to 1 and pretty well
represent the span of apparent sizes of all planets except Pluto. As you can
see, this last planet is far removed from the rest in apparent size, and is in
fact far closer to Betelgeuse, one of the largest stars in actual as well as
apparent size. This suggests that generalizations like "stars are point
sources and planets are extended objects" can be misleading if taken
literally; neither Pluto nor Betelgeuse is an extended object to amateur
equipment, while both can be photographed with perceptible disks by the
largest observatory instruments.

Visual resolution is the ability to separate nearby points into distinct
visual images, equivalent to the ability to see a very small object as having
a discernible size. The figure 4′ for visual resolution is a very rough
estimate varying with many factors, starting with the individual's visual
acuity. But the value does indicate why we have little trouble resolving
Alcor-Mizar without optical aid, but are unable to see the planets as disks
with the naked eye.

Dubhe — Polaris	29°
Alkaid — Dubhe	26°
Deneb — Albireo	22°
Betelgeuse — Rigel	18.6°
Merak — Dubhe	5.4°
Alnitak — Mintaka	2.7°
M31 (length)	2.7°
M42	1.1°
Moon	31'
M13	23'
M51 (length)	12'
Mizar — Alcor	12'
Visual Resolution (approx)	4'
M57	1'
Jupiter (at maximum)	48"
Neptune	2.5"
Pluto (approx)	0.1"
Betelgeuse	0.05"
Diffraction Disk (approx)	r/750 (mm)
Film Resolution (approx)	0.02 (mm)

Table 3-1. The Measure of Common Celestial Objects, Lens Diffraction Disk and Film Resolution.

At the bottom of the list are two items that elucidate a point made in Chapter 2 about the lower limit on image size. Film resolution is dependent on the size of the emulsion grain — the minimum bits of emulsion that can record pieces of the image — and varies greatly with different types of film. The diffraction disk is the smallest possible image of a point source that can be formed by a bundle of parallel light rays passing through an aperture, as a result of the nature of light itself, totally independent of any additional enlargement due to the lens. In other words, it is the image of a true point at an infinite distance, formed by a lens with no aberrations. As we shall see, such lenses are not that common; but, even if we had one, it would still image our point sources no better than measurable disks, with diameters directly proportional to focal ratio. In real life, however, film resolution and diffraction disks tend to be the least of our problems. Film also contributes something called halation, which is the enlargement of brighter images by light scattering among the particles of the emulsion. Lenses, as noted, enlarge images as a result of various aberrations (discussed at length in Chapter 4). Worst of all, atmospheric turbulence also adds to the problem, as that very poetic "twinkling" translates directly into image motion across the surface of our film.

The net result of all this is that 0.1 mm turns out to be a pretty good value for our lower limit on image size. Therefore, when our formulas start predicting image sizes below that 0.1 mm threshold, it's time for a bit of healthy skepticism — and probably time to forget about image size formulas and start treating your subject as a point source.

IMAGE SIZE

Given a subject of apparent angular size A, and a distortion-free lens of focal length f, the precise formula for image size s is:

$$s = 2f \left(\tan \frac{A}{2} \right)$$

Focal length and image size can be in any units, using the same for both. Millimeters are conventional and should be assumed when not specified.

When subject size A is expressed in degrees, a very good approximation for image size s is:

$$s \approx \frac{fA}{57.3}$$

For A less than 20°, this approximation will differ from the precise value by no more than 1%, becoming much better than that with smaller subjects. Consequently, it is used far more frequently than the precise formula given above. It tells us, for example, that our 50 mm normal lens will put the 22° Northern Cross onto about 19 mm of film.

The derivations may be found in Appendix C.

FORMAT COVERAGE

Assuming that the available image area of a mounted 35 mm slide is 22.9 mm by 34.2 mm, and that our lens has a focal length of f mm, our slide format will cover an area of sky that is precisely:

$$2 \arctan \frac{11.45}{f} \quad \text{by} \quad 2 \arctan \frac{17.1}{f}$$

A good approximation of this, in degrees, is:

$$\frac{1312}{f} \quad \text{by} \quad \frac{1960}{f}$$

For f greater than 100, i.e., focal lengths in excess of 100 mm, these values will be less than 1% off the precise ones, and will improve with increasing focal lengths. For example, our 135 mm telephoto will cover about 9.7° by 14.5° of sky. However, as we get down toward the so-called normal 50 mm, the approximation starts getting farther off the mark; it is totally useless in trying to determine the coverage of wide-angle lenses.

The (trivial) derivation is in Appendix C.

Trailing

Although we were taught that the earth rotates within a relatively stationary universe, we still persist in discussing the situation in classical geocentric terms. Thus, we check to see what time the sun will "rise" or "set," and we speak of the stars' diurnal "motion." For our purposes, it really makes no difference, so we'll sacrifice pedantic precision in favor of a more casual approach. As a result of this motion, when we try photographing the sky with a stationary camera (e.g., a camera mounted on a regular tripod), the stars tend to record as trails across our film, rather than images of points. Extended objects like the moon tend to "smear" rather than form clear images. This phenomenon is called trailing. As mentioned earlier, this problem goes away entirely with the use of a properly aligned astrophoto mounting, but that's a subject for Chapter 6. For now, let's look at a much simpler attack. As most photographers know, one answer to subject motion — or camera motion, since it doesn't matter whether you're photographing an airplane from the airport or photographing the airport from an airplane — is higher shutter speeds, i.e., shorter exposure durations. In keeping with things astrophotographic, our "shorter exposures" would probably be considered rather long by the average photographer.

Since the stars rotate about the celestial poles, they do not all move at the same speed. Only the stars on the celestial equator travel 360 real full-size degrees in each daily rotation; all others travel in smaller circles, with correspondingly smaller degrees. As an extreme case, Polaris, less than one degree off the actual pole, completes a full circuit around that pole each day; however, the circumference of that circle is only about six full-size degrees. Hence, equatorial stars move along at a fair rate, while Polaris barely moves at all. The key, of course, is the star's declination — i.e., distance from the celestial equator — measured in degrees and generally represented by the lower case greek letter "delta", δ.

To quantify all of this, let us again assume a lens of focal length f. We will also assume a subject at declination δ, and an exposure duration of t seconds. Letting d represent our subject's image drift or trail length on our film, the precise formula is:

$$d = 2f \tan \frac{t \cos \delta}{480}$$

where d will be in the same units as f, generally millimeters. For t less than 4800 seconds, i.e., exposure durations under 80 minutes, a very good approximation, giving values no more than 1% off the precise ones, is:

$$d \approx \frac{f\,t}{13750} \cos \delta$$

A further simplification occurs if we assume equatorial stars, i.e., worst case:

$$d \approx \frac{ft}{13750}$$

We can then adjust if we wish by multiplying our resulting d by 3/4 for stars around $\delta = 40°$, 1/2 for stars at $\delta = 60°$, and 1/4 for stars around $\delta = 75°$.

On the other hand, if we wish to know how long to expose our film to obtain a specified trail length d, we can use the approximation:

$$t \approx \frac{13750\,d}{f \cos \delta}$$

This will give results less than 1% from the exact as long as d does not exceed $f/3$. As before, if we start by assuming equatorial stars, we can use the simpler:

$$t \approx 13750 \frac{d}{f}$$

We could then adjust by multiplying the resulting t by 4/3 for stars around $\delta = 40°$, 2 for stars at $\delta = 60°$ and 4 for stars around $\delta = 75°$. Derivations are in Appendix C.

With these formulas, we can now predict trail lengths for specified exposures and/or durations to obtain specified trail lengths. But what if we don't want any trailing at all? Why, we simply do as any other photographer does who wants to "stop" motion: use the aforementioned shorter exposure. Recalling that point sources don't actually image as points anyway, what we do is simply pick a trail length so small as to be indistinguishable from the image of a point.

s ≡ size of image at focal plane, in mm†
d ≡ amount of image drift due to earth's rotation, in mm†
A ≡ apparent angular size of subject, in degrees
δ ≡ declination of subject, in degrees
f ≡ focal length of lens, in mm†
t ≡ duration of exposure, in seconds

$$s = 2f\left(\tan\frac{A}{2}\right) = f\,\frac{A}{57.3}\ \text{within 1\% when A} < 20°$$

coverage on 22.9 by 34.2 slide format:

$$2\arctan\frac{11.45}{f}\ \text{by}\ 2\arctan\frac{17.1}{f}$$

$$\frac{1312}{f}\ \text{by}\ \frac{1960}{f}\ \text{within 1\% for } f > 100\ \text{mm}$$

$$d = 2f\tan\left(\frac{t\cos\delta}{480}\right) = f\left(\frac{t\cos\delta}{13750}\right)$$

$$= f\,\frac{t}{13750}\ \text{within 1\% for t} < 4800\ \text{arc seconds at } \delta = 0\ \text{(worst case)}$$

$$t = \frac{480\arctan(d/2f)}{\cos\delta} = \frac{13750\,d}{f\cos\delta}$$

$$= 13750\,\frac{d}{f}\ \text{within 1\% when d} < f/3\ \text{at } \delta = 0\ \text{(worst case)}$$

†Note that s, d, and f need not be in mm, as long as they are consistent; convention suggests the use of mm.

Table 3-2. Image Size, Format Coverage, and Image Trailing

Focal Length in mm	Coverage on 35 mm	Image Size in mm for Subject Angle:							Duration in Sec for 0.05 mm Trail at Declination:		
		15"	1'	4'	15'	1°	4°	15°	0°	50°	75°
28	44° × 63°				0.12	0.49	2.0	7.4	25	38	95
35	36° × 52°				0.15	0.61	2.4	9.2	20	31	76
50	26° × 38°				0.22	0.87	3.5	13	14	21	53
85	15° × 23°				0.37	1.5	5.9	22	8.1	13	31
135	9.7° × 14°			0.16	0.59	2.4	9.4		5.1	7.9	20
200	6.6° × 9.8°			0.23	0.87	3.5	14		3.4	5.3	13
300	4.4° × 6.5°			0.35	1.3	5.2	21		2.3	3.6	8.9
400	3.3° × 4.9°		0.12	0.47	1.7	7.0	28†		1.7	2.7	6.6
500	2.6° × 3.9°		0.15	0.58	2.2	8.7			1.4	2.1	5.3
750	1.7° × 2.6°		0.22	0.87	3.3	13			0.92	1.4	3.5
1,000	1.3° × 2.0°		0.29	1.2	4.4	17			0.69	1.1	2.7
1,500	52' × 78'	0.11	0.44	1.7	6.5	26†			0.46	0.71	1.8
2,000	39' × 59'	0.15	0.58	2.3	8.7				0.34	0.53	1.3
3,500	22' × 34'	0.25	1.0	4.1	15				0.20	0.31	0.76
10,000	7' × 12'	0.73	2.9	12					0.07	0.11	0.27
25,000	3.1' × 4.7'	1.8	7.3	29†					0.03	0.04	0.11

Table 3-3. Focal Length Implications. All values are approximate. Image sizes below 0.1 mm are omitted as being unrealistic. Image sizes above 34.2 mm are omitted since they are too large for the standard 35 mm film frame length while those above 22.9 (marked †) are too large for the frame width.

For all practical purposes, a good choice is 0.05 mm — half our "threshold" size. For stars (or, for that matter, any subjects) at or near the celestial equator, and a lens of focal length f mm, all we need do is keep our exposure durations (in seconds) less than $700/f$. A stationary camera with a lens of f mm focal length will produce essentially trail-free images on exposures under $700/f$ seconds. This is a rule that we will refer to often. For the fussy, reducing that to $500/f$ will reduce trail lengths to about 0.036 mm, which begins to approach the neighborhood of the film's resolution, essentially eliminating trailing altogether.

The reasoning behind these limiting values is simply that while bloated images from overexposed bright stars can be considerably larger, good crisp star images should probably run to about 0.1 mm in diameter. How much can this be elongated by trailing, and still look acceptable? The amount of elongation of an image is conveniently expressed as its

	Aperture in cm	Focal Length in mm	Focal Ratio
Average Human Eye (approx.)			
wide open	0.7	17	2.4
stopped-down	0.2	17	9.5
Some Typical Lenses for 35 mm Cameras			
Ultra-Wide (21 mm or less)	0.5	17	4
Wide-Angle (24 to 35 mm)	1	28	2.8
Normal (40 to 60 mm)	3.5	50	1.4
Short Telephoto (85 to 180 mm)	5	135	2.8
Medium Telephoto (200 to 400 mm)	7.5	300	4.0
Long Telephoto (500 to 1200 mm)	14	800	5.6
A Common Schmidt Camera	14	225	1.65
A Basic 6″ Newtonian	15	1,200	8
A Common 8″ Schmidt-Cassegrain	20	2,000	10
Palomar Mountain Schmidt 48″	120	3,000	2.5
Mount Wilson 60″ Reflector	150	7,600	5
		24,000	16
		46,000	30
Mount Wilson Hooker (100″)	250	12,800	5
		40,000	16
		75,000	30
Palomar Mountain Hale (200″)	500	16,800	3.3
		80,000	16
		150,000	30

Table 3-4. Comparative Optics

aspect ratio, which is simply the ratio of its shorter dimension to its longer one. Let's adopt, perhaps somewhat arbitrarily, a limiting aspect ratio of 2:3 — i.e., dimensions in the ratio 2 to 3 — which means a size of 0.1 by 0.15 mm. This implies a 0.05 mm limit on trail size, leading to the $700/f$ limitation on exposure duration. The $500/f$ suggestion "for the fussy," means a trail length of about 0.036 mm, giving images of 0.1 by 0.136 —

or an aspect ratio of about 3:4, which should be good enough for anyone. Note that these are aspect ratios for crisp star images; for the oversize ones, the effects of trailing become relatively smaller.

Of course, we can still take advantage of slower targets away from the equator by increasing our exposure times, either by dividing by cos δ or by multiplying by our adjustment factor for the appropriate declination.

Table 3-2 recaps the formulas developed in this chapter, while Table 3-3 illustrates these formulas for a variety of focal length values. Table 3-4 summarizes the wide range of optics in astrophoto use.

Congratulations! You have now gone through all the theory we need. From here on, we start putting it all to use, getting down to the nuts and bolts of how to photograph the sky.

CHAPTER 3 EXERCISES

The answers to these exercises are given, and explained, at the back of the book.

1. What would be the image size for a 26° subject — e.g., the Big Dipper — photographed with a 50 mm lens?

2. What is the image size of a 2.7° subject — e.g., Orion's belt — produced by a 135 mm lens?

3. Assuming a 31′ subject — e.g., the moon — what focal length is required to produce a 5 mm image?

4. To get a 3 mm image, what focal length would be required for a 48″ subject — e.g., Jupiter at its largest?

Chapter 4
BASIC EQUIPMENT

As mentioned in Chapter 1, a great deal of astrophotography is possible with relatively modest equipment. What we need, at the absolute minimum, is some sort of camera body, a lens, and a support. Since some equipment is better than other equipment it is worthwhile to research before purchasing.

THE NOTEBOOK

As for the notebook, any one you use conscientiously is a good one. That might even be small scraps of paper on which you scribble your data until it can be transferred to the slide mount. Why bother? You don't need a notebook to take an astrophotograph, but you will be glad to have it when you want to rephotograph a subject with a new piece of equipment, or perhaps simply to do it better. You will also find it extremely useful when photographing a new subject to refer to data you already have on a similar subject. Remember, the book you are now reading will give you guidelines, recommendations for starting points for various subjects, but the refinement and perfection of your technique is your responsibility. You can document your efforts and learn from them, or you can hope for the best each time you start essentially from scratch.

THE CAMERA

Almost any kind of camera will do, but many cameras are extremely limited in capabilities, and therefore limiting to the photographer. Without question, the camera of choice for the vast majority of astrophotographers, and would-be astrophotographers, is the 35 mm single-lens reflex. It is unmatched for convenience and versatility. The single-lens reflex, or SLR, has a single lens both for viewing/focussing and for the actual photographing; it uses a mirror to reflect the viewing/focussing image to a location away from the film. The SLR designation is very descriptive, but, in another sense, very misleading. A great virtue of viewing and focussing your subject through the very lens with which you will photograph it is that you can easily replace that lens with any one of a vast number. In other words, while the SLR uses only a single lens at any one time, it is equally appropriate to think of it as a many-lens reflex. It is this ability to function with a variety of lenses that endears the SLR to the astrophotographer.

Photograph 4-1. A more-or-less typical SLR with its "normal" 50 mm lens.

It is clear that, in astrophotography, the 35 mm SLR is vastly superior to all other cameras. But that does not mean that all 35 mm SLR's are created equal. They are not. Let's see what else we should consider in selecting our astrocamera. Since we know we will want to put our camera on a tripod, a tripod socket is essential. This should be no problem; few decently made cameras lack this feature. The other essential is full manual control of exposure, including the ability to keep the shutter open for protracted periods. This still leaves us many, many candidates, but it does rule out some of the newer, more automatic models.

Any well-made 35 mm SLR with the above attributes will do just fine. But, if we want to be really fussy, there are three other things we could look for. First is a provision for separately raising, or "locking," the viewing mirror a few moments prior to exposure. Under certain conditions, this conveniently reduces camera vibration, but is certainly not essential. The second is a removable pentaprism, (that lump atop the camera that you look into). With interchangeable finders, the standard pentaprism can be replaced by a finder that lets you look directly down at the viewing screen, simultaneously offering a magnified brighter viewing/focussing image as well as a more comfortable working position. Lacking this ability, most cameras offer an add-on right-angle accessory which attaches to the fixed pentaprism. While not suitable for all subjects, due to dimness of image, it does offer an improved viewing posture. But be warned: replacement or add-on, many "down-looking" viewfinders provide images that are left-to-right reversed—one more source of potential confusion. Third and last is the ability to interchange the camera's viewing screen—the surface between reflex mirror and pentaprism on which the viewing/focussing image is formed. Those that come with the camera usually have some combination of ground glass plus micro-prism and/or split-image focussing aid. These are fine for most photography, but less so for dim light and/or slow lenses. The ability to substitute a screen more compatible with our needs—such as very fine ground glass with nothing added—is offered by some SLRs, and can make our task a bit easier, especially when tackling the more challenging subjects.

LENSES

As usual, we use the term "lens" as short-hand for image-forming optics—lenses, mirrors, or combinations thereof. As discussed in Chapters 2 and 3, we are very concerned with lens focal length and aperture, wanting a great variety in the first and as much as possible of the second. But lenses have other attributes. Until now, we have dealt with lenses as though they were perfect, i.e., as though they formed a diffraction disk of the appropriate theoretical size for each point in our subject. Such optics do exist; they are described as diffraction-limited, meaning they image point sources as small as the nature of light allows. In other words, none of the ills lenses are prey to are permitted to smear the image beyond the irreducible minimum size. These ills are known collectively as aberrations, and there is an impressive variety of them. They affect all photography but they affect astrophotography more.

If you photograph a face, for example, the fact that point images are actually blurs is not usually a problem. One point on a cheek looks much like any other point on a cheek, so the light lost from the image of one point is gained from the image of its neighbor. This all balances out and becomes invisible, except when we come to an edge. Where eye meets lid, the light-swapping between eye points is still fine, as is the light-swapping between lid points. But the light-swapping across the boundary causes a softening of the image to a degree, depending on the seriousness of the aberrations present. But that is all: merely the loss of some "snap" in the image. In astrophotography, the situation is worse. Since our pictures consist largely of point images, there is not the light-swapping present in other photos. Aberrations cause light from the star images to spread into the otherwise dark sky image: it's a one-way process. As though it weren't bad enough that we can't image stars as points, now we can't even image them as small disks. What we get is odd-shaped blobs of light, which tend to be larger farther from the center of the frame. For this reason it is standard practice to test lenses on an optical bench using "artificial stars." If a lens images a star well, it will image everything well; if there are aberrations beyond the diffraction disk, they will be revealed by their characteristic shapes.

The major aberrations are as follows.

1. *Astigmatism* is very common in fast camera lenses used at full aperture, e.g., around $f/2$ or faster. It causes star images to be pulled out of shape, into smears running toward and away from, or in arcs around, frame center, or a combination of the two. The effect is aggravated with distance from frame center.

2. *Coma* turns star images into comet-like images with tails pointing away from frame center, and is aggravated with increasing distance from frame center. It can occur in any lens, but is inevitable in parabolic mirrors such as those used in reflecting telescopes. In fact, in such mirrors, the "coma-free field," i.e., the image area in which discernible coma is smaller than the diffraction disk, has a diameter in millimeters roughly equal to the square of the focal ratio.

3. *Spherical Aberration* results from the fact that a spherical surface does not have a constant focal length for all light rays striking it. That is why reflecting telescopes use parabolic mirrors (see coma, above).

4. *Chromatic Aberration* causes colored fringes around bright images, resulting from different colors of light coming to focus at different focal lengths. It occurs only with lenses, not mirrors, since it is a refraction aberration only.

5. *Distortion* is the rendering of straight lines in the subject as curves in the image, from a systematic displacement of image points. It is classified as pincushion or barrel distortion, respectively, for lines that bow in toward center or bow out away from center. The fish-eye lens has deliberate extreme barrel distortion. This aberration affects the astrophotographer less than the terraphotographer, especially the architectural photographer. Mild distortion is not a serious problem for the astrophotographer, except in astrometry, i.e., the determination of stellar positions by precise measurements of photographic plates.

We must also be on guard against two other ills that, strictly speaking, are not aberrations. The first is vignetting. When this occurs, the outer portions of the frame receive less illumination than the center. It can be an abrupt cutoff giving a circular image with a well-defined edge, or it can be a gradual fall-off becoming worse as we approach the frame corners. The latter is very common, particularly with fast camera lenses at full aperture and also with reflecting systems with undersized secondary mirrors. Finally, there is curvature of field, which means that the focal surface, i.e., the location in space of the image of sharpest focus, is not a plane but some part of a curve — frequently a sphere. This means that, as long as the film is held in a plane, only part of the photograph can be in focus, usually the center.

Photograph 4-2. An assortment of SLR lenses; focal lengths are 28, 50, 85, 135, and 300 mm.

This is far from an exhaustive list of optical gremlins, but it does include the most common. The real lesson to be learned here is that just because your lens does a splendid job in general photography, don't assume it will do as well on the sky. As indicated, stars are the ultimate test of lens quality. Given the option, use the best. When a less-than-ideal lens must be used, try stopping down somewhat and compensating with longer exposures with perhaps slightly greater trailing, or else use faster film with its slightly coarser grain. Experiment to find the trade-off that works best for you. Don't forget to use that notebook!

Photograph 4-3. An SLR body "close-coupled" at the prime focus of a Schmidt-Cassegrain telescope.

COUPLING HARDWARE

One problem we have ignored is the matter of attaching camera body and lens. For the most part, each brand of camera has its own unique lens mount, incompatible with all other brands. There is no problem if you stick to lenses made by the camera-maker: Whizzar lenses built specifically for the Whizzon SLR body. Also, many independent lens-makers produce lenses that fit most camera bodies. The real problem arises when you want to attach camera body and telescope, neither designed with the other in mind. Why would you wish to do that anyway? Well, it just happens that a telescope objective — the larger lens at the sky end of the optical path — forms a real image exactly the way a camera lens does. Normally, this image is examined visually via the telescope eyepiece — the smaller lens at the eye end of the optical path, which is essentially a high-quality magnifying glass. The image formed by the objective can be recorded directly on film, simply by removing the eyepiece and replacing it with a lensless camera body. Thus, the telescope sans eyepiece effectively becomes a long-focus camera lens. This can be done by a T-ring or T-adapter, a device invented years ago to enable a single lens to fit a variety of camera bodies. The principle is elegantly simple. Instead of putting Whizzar hardware on the camera end of the lens, restricting it to Whizzon bodies only, the lens-maker simply finished his lens with a standardized male thread called, for no obvious reason, a T-thread. Then, a series of adapter rings was provided, each with a different camera fitting at the back and a female T-thread at the front.

Screwing the appropriate ring onto the rear of the lens instantly adapted it to the camera of your choice. The system has its limitations, but is still in fairly wide use for lenses that lack today's sophisticated automatic functions.

The other half of the problem was also solved many years ago, by telescope makers. Telescopes have a standard eyepiece barrel diameter: 1-1/4 inch (standard but not universal). Thus, any 1-1/4-inch tube can be inserted and locked into most telescopes. If such a tube carried a male T-thread on one end, it could accept a T-ring and thus a camera body. That is precisely what is done. To the telescope, it is much like attaching a very heavy eyepiece; to the camera, it is like attaching a very heavy lens. Certain telescope manufacturers offer a slight variation on their telescopes: the eyepiece holder itself is replaced with a special camera coupler carrying the T-thread. Regardless of the details the result is effectively a 35 mm SLR with a very powerful telephoto lens.

Photograph 4-4. The Basic Equipment: an SLR body with 300 mm lens mounted on a (sturdy) tripod. Note that at focal lengths of 300 mm and up, it's the lens — not the camera body — that is attached to the tripod.

THE TRIPOD

Long-focus lenses produce magnified images: the greater the focal length, the greater the magnification. Unfortunately, these lenses are equal-opportunity magnifiers: in addition to the image of your subject, the effects of camera motion will also be magnified in direct proportion to lens focal length. This is the reason for the rule of thumb mentioned in Chapter 2: hand-holding a lens of f mm focal length, use shutter speeds no lower than $1/f$—e.g., 1/30 for 28 mm, 1/60 for 50 mm, 1/125 for 135 mm, 1/250 for 200 mm, 1/500 for 500 mm, 1/1000 for 1000 mm. By now, we should be pretty well resigned to the fact that, in astrophotography, long lenses are common and short exposures are anything but. The hand-held astrophotograph is a rarity.

Note that, as used here, the term "tripod" is short for "good solid tripod," which should be understood throughout. Photograph 4-5a shows a lens test chart, taken with a 1500 mm lens, long enough to be a good test of a tripod's mettle. Exposure duration was 1/15 second, about as bad as can be. Short exposures, e.g., 1/500, tend to freeze motion, including camera shake. Long exposures, e.g., five minutes, record relatively little at the beginning while the camera is still vibrating from mirror and shutter action. But 1/15 sec will record all that vibration very nicely. As you can see, there was plenty to record. The tripod used was a good one, not one of those useless pocketable flimsies with six-section legs, but a reputable four-pound model with two-section legs. It is perfectly adequate for many applications, but not 1500 mm at 1/15 sec.

Photograph 4-5b has exactly the same setup: 1500 mm at 1/15 sec. However, the tripod was a ten-pound studio model costing as much as a moderate-priced telephoto lens. The result is certainly a great improvement, but notice that the verticals and horizontals did not fare equally well: it looks as though something had bounced the camera up and down. That is because something bounced the camera up and down. The something was our reflex mirror. Recall that we mentioned a mirror that could be raised prior to exposure as a desirable feature. This photograph illustrates that need.

Photograph 4-5c finishes the story of mirror lock-up. Except for locking the mirror, everything is identical to the previous exposure. Although improved further still, verticals and horizontals are again not equal. This time, it looks as though something kicked the camera sideways. You guessed it: something did. The culprit was the shutter itself, which travels horizontally on this particular camera, as it does on many. The solution to that problem is the old hat trick, the preferred technique for exposures of several seconds or longer. Being careful not to actually touch the telescope or lens, cover it by holding a hat or dark card or the like just in front of it. Then open the shutter, wait a few seconds for the

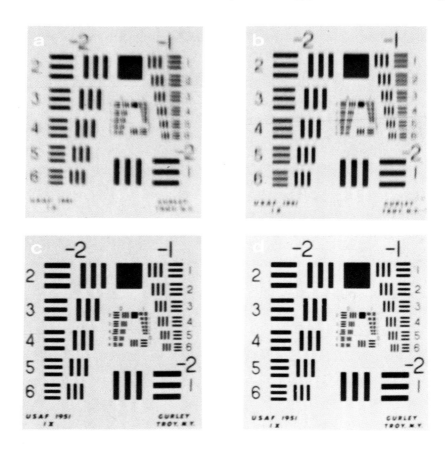

Photograph 4-5. Tripod tests, 1500 mm, 1/15 sec. Frame a: a good, medium-duty 4-pound tripod, mirror-free. Frame b: a very solid 10-pound tripod, mirror free. Frame c: the same ten pound tripod, but mirror locked before exposure. Frame d: a 25-pound telescope tripod, mirror locked.

vibrations to die out, and remove the hat or whatever to actually start the exposure. This obviously cures the mirror bounce problem as well, which is why the lock-up feature was described as being handy only on occasion. That occasion does arise as we do use shutter speeds of 1/15, 1/30 — we use them all, so we can't use hats instead of shutters all the time.

But photograph 4-5d shows the real solution. Again 1500 mm; 1/15 sec. This time, however, the ten-pound studio tripod with two-section legs has been replaced by a fifteen-pound telescope tripod with solid legs. Note that a good solid support is essential even if the hat trick is used. Camera mechanism is not the only source of vibration; it was simply the most

convenient source for these tests. In the field, wind and the photographer
will prove much greater hazards. The lesson is clear: hat trick or shutter,
use the most solid support you can.

A word on basic auxiliary equipment: your tripod's inseparable com-
panion should be a cable release. This is a flexible cable that screws into a
special socket in the camera—usually the shutter release button; by a
button at the other end, the shutter can be tripped without actually
touching the camera. These come in various lengths. You want one that is
long enough to isolate the camera from any motion at your end, but not
so long as to get in the way. Many are available with setscrews or other
locking mechanism to keep them in the "pushed" position (thus holding
the shutter open) until released; this is a must for the astrophotographer.
Several tripods have a provision for attaching the cable release to one of
their control handles, enabling the user to press the release while holding
the handle. My recommendation is don't use it! Camera motion induced
by pushing the tripod has no readily discernible virtues over camera
motion induced by pushing the shutter button.

Chapter 5
THE EASY SUBJECTS

The variety of subjects available to our basic equipment might surprise you. Let's look into them and see how each can be photographed.

STAR TRAILS AND METEOR TRAILS

We will start with perhaps the simplest subject of all. Select a lens to cover the area you want, load a suitable film, set up your trusty tripod with camera aimed in the proper direction, check for lens focus at "infinity," set the aperture (generally the widest available as we're not producing point images anyway), check for "B" shutter setting, and go. What you will get is trails of stars in the photographed field. The lengths of those trails depend on exposure duration; lengths also depend on distance from the celestial equator, as does curvature. The real surprise for the newcomer is the amount and variety of color in the stars, most of which is lost to the naked eye due to the low light levels.

Occasionally, a meteor will flash across the field and leave its trail on our film. This will be easily distinguished as a straight trail cutting across the curved ones; also, meteor trails frequently vary in brightness along their length, revealed by the varying trail width in our photograph. In sharp contrast to the vast majority of astrophotography, there is precious little planning to be done for meteor photography; we can't even predict their appearance, much less their size or brightness. We can improve our chances by taking advantage of predicted meteor showers, in which relatively large numbers of meteors appear over a few days, but by and large meteor photography is very much a matter of luck.

Getting back to the star trails, what can we expect to capture on our film? As you now know, that depends on a number of things, primarily film speed and lens speed (in absolute aperture, since we are photographing point sources). Thus an ISO 200 film will show fainter stars than an ISO 64 film; but a 200 mm $f/4$ lens will show fainter stars than a 50 mm $f/1.4$ lens, since the 200/4 aperture is 5 cm while the 50/1.4 aperture is under 3.6 cm.

The real catch in this setup is exposure duration. Remember that the stars' images are trailing, so the image of each star remains on a given spot of film for a very limited time, regardless of how long the shutter is open. As discussed in Chapter 3, that period of time is $500/f$ for stars near the equator, increasing as we move toward the poles. Knowing your lens focal length, you can determine the effective exposure duration which is less than the total exposure duration. This is true for stars, but bear in mind we are also photographing that most extended of all extended objects, the sky. Here, each point looks just like its neighbor, so the exposure of the sky does build up, and its effective exposure duration is the entire time the shutter is open. Further, the sky is not completely dark: we have miserable polluted skies and good dark skies, but never completely dark skies. After a while, the sky background in our photo will become bright enough to blot out the fainter star trails. This happens sooner with faster film and/or faster lenses, this time faster in relative aperture. So our ISO 200 film produces light skies sooner than our ISO 64 film, thus undoing to an extent its good work in capturing fainter stars. But our 200/4 lens is still outdoing the 50/1.4, not only recording fainter stars with its faster absolute aperture, but also keeping the sky background darker by virtue of its slower relative aperture. This idea that a slow relative aperture can be a plus is unique to astrophotography. But it is a very important idea, which we will treat at length in Chapter 7.

THE MOON

Astrophoto subject #1, the moon is also an appropriate subject for the gag about the good news and bad news. Starting with the good news: although we usually view it at night, the moon is essentially a daylight subject—a piece of sunlit rock. As a result, exposures for the moon are not all that different from other daylight exposures, e.g., landscapes, here on earth. However, as the moon goes through its phases, we must adjust those exposures—just as we do here on earth for side-lighted or back-lighted subjects. In the bad old days, aspiring photographers were told early on to be sure that the sun was at their backs—i.e., to have their subjects front-lighted—since a front-lighted subject does its best job of reflecting incident light. Today—with the considerably greater latitude

Photograph 5-1. Star Trails over Kitt Peak, taken March 1979. 50 mm $f/1.8$ on a fixed tripod using High Speed Ektachrome push processed to ISO 400. Duration 30 minutes, giving total fx of 36 for the sky, but effective fx' for stars of only $13'$ due to trailing. By David Healy.

offered by our faster lenses and emulsions—the rule-of-thumb is simply: open up one stop for side-lighting, two stops for back-lighting. That rule-of-thumb doesn't quite apply to the moon, the moon's "complexion" being unusually rough, but Table 5-1 summarizes exposure recommendations for the moon in its various phases:

	fx
Full Moon	10 to 12
Gibbous Moon	11 to 13
Quarter Moon	13 to 14
Crescent Moon	14 to 15
Area Lighted by Earthshine	24 to 25

Table 5-1. fx exposure values for moon in its various phases.

The column headed fx is the recommended total fx in the effective exposure system. This is the standard form in which exposure recommendations will be given throughout the book. Again, as discussed in Chapter 2, correct exposure is a myth, and even if there were such a thing, it would remain beyond the power of any systematic approach. In some cases, including this one, there is so much variation as to preclude offering a single firm recommendation, so we shall show the consensus range. Part of the variation here results from atmospheric extinction (discussed in Chapter 2). That explains why the range is greater for the full and gibbous phases, which can vary greatly in altitude, as compared to the others, which tend to be visible only lower in the sky.

But remember: whatever values are shown, they are not "correct" values; they are merely the best consensus recommendations available. Thus, when feasible, it is advisable to bracket, that is to take exposures on either side of those recommended.

Now for the bad news: image size. Imposing as it often looks in the night sky, the moon's apparent diameter averages just about 31′ of arc—just over half a degree. That includes the huge orange full moon sitting just above the horizon; the hugeness is strictly a psychological illusion that is called, oddly enough, the moon illusion. The actual variation in the moon's apparent size, resulting from the variation in its distance from earth, is only about 4′ of arc. If we put 31′ or rather 31/60° into our image size formula, we find that we can approximate the moon's image size fairly well by using $f/111$. To recall just what that means, take another look at Photographs 2-1, which show how the moon looks on our 35 mm frame when photographed at focal lengths of 50, 135, 300, 750, and 1500 mm. But cheer up; at least (barring the very widest of wide-angle lenses) it always remains an extended object.

BRIGHTER STAR FIELDS AND CONSTELLATIONS

Wide-field sky photography is easy and can provide truly rewarding results. As with star trail photos, a wealth of stellar color can be captured on film, along with the brightness variation and striking patterns. The technique is similar to that discussed earlier for star trails, except that we now limit our total exposure duration to prevent visible trailing. The limit, of course, is $700/f$ seconds (or else $500/f$, if we wish to be demanding) for the neighborhood of the celestial equator and suitably adjusted for elsewhere, as discussed in Chapter 3.

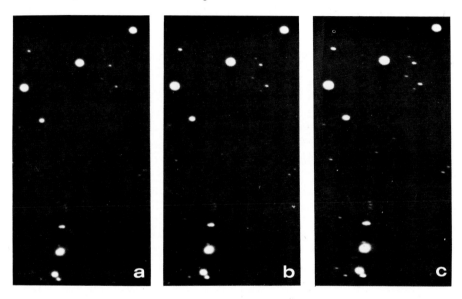

Photograph 5-2. Orion's belt and sword, taken January 1981 in Everglades National Park, FL. All exposures are 50 mm $f/1.4$ on Ektachrome 400 film, using a fixed tripod. Durations, in order, are: 10, 14, and 20 sec — or $500/f$, $700/f$, and $1000/f$. Note that, in Frame c, the useful effective exposure for each star image was rather less than 20 sec, due to trailing. For frame a, the total $fx' = 14'$.

Precise exposure data here is totally out of the question. But let's assume that any exposure will be pegged at a specific magnitude. This means that brighter stars will record on our film, but they will be overexposed with artificially enlarged images and washed-out colors. Likewise, we shall also record dimmer stars, probably about three-plus magnitudes dimmer, but increasingly faint and difficult to discern. In all cases, the range of recorded magnitudes and the image degradation for each will depend upon film latitude and other factors far too complex to pin down. We shall therefore confine ourselves to considering the optimum or central magnitude to which our exposure is pegged. Representing the fx value for that magnitude by fx' — to remind us that it refers to a point source, a convention we shall use from here on — our guideline rough approximation for stellar exposures is:

$$fx' \approx \frac{4}{3} \text{ magnitude} + 9$$

This is derived in Appendix D.

In lieu of the formula, Table 5-2 should prove more than adequate. Note that, like contest prizes and league standings, greater magnitudes are denoted by lower numbers; however, magnitudes go beyond the first, to magnitude zero and even to negative magnitudes for the brightest objects.

Magnitude	fx'
−3.0	5
−1.5	7
0.0	9
1.5	11
3.0	13
4.5	15
6.0	17
7.5	19
9.0	21
10.5	23
12.0	25
13.5	27
15.0	29
16.5	31
18.0	33

Table 5-2. fx' exposure values for stars (point sources) based on stellar magnitudes.

For starters, fx' in the 10 to 12 range will center exposure around magnitudes 1 to 2. Assuming that film latitude gives us another three magnitudes, we could expect images of stars down to the magnitude 4 to 5 range. This approximates the grasp of the unaided eye under average skies, but will show much more color than the naked eye can see. The results can be beautiful. But don't be misled: this is not the same 10 to 12 fx range we suggested for the full moon; this is a very different fx'.

Photograph 5-3. The constellation Taurus, taken January 1981 in Everglades National Park, FL. 50 mm $f/1.4$ on Ektachrome 400 film, using a fixed tripod. Duration 10 sec, or $500/f$. Total $fx' = 14'$.

Just to nail that down, let's try an example, assuming ISO 64 film and a 200 mm $f/4$ lens. Our film, of course, contributes an fx of 2, whatever we are photographing. For the moon, our lens gives us 6 more toward the $fx = 10$ we need. The remaining 2 comes from an exposure duration of 1/1000 of a second. (As we said, the moon is really a daylight object.) For star fields, on the other hand, our 200/4 = 50 mm = 5 cm aperture gives us a -5 fx' contribution toward the needed $fx' = 10$. Our exposure duration must therefore contribute a whopping 13 to make up the deficit, calling for a full 2-second exposure. Fortunately, in this example, $700/f = 700/200 = 3.5$, so we are still able to expose without trails; in fact, since we are within the stringent $500/f$ limit, we can proceed with supreme confidence. To repeat: for point sources, you need raw aperture; fast focal ratios buy you nothing.

Although we've talked only about stars in this section, we deal with planets the same way; we use their magnitudes and corresponding fx' values when using lenses too short to resolve them into extended objects (discussed further in Chapter 9). Although the magnitudes of the planets are variable, some of them dramatically so depending on their locations with respect to earth and sun, Table 5-3 shows the general expected

Photograph 5-4. The constellation Cassiopeia, taken January 1981 in Everglades National Park, FL. 50 mm $f/1.4$ on Ektachrome 400 film, using a fixed tripod. Duration 28 sec, or $1400/f$—twice as long as we can get away with near the celestial equator. Total $fx' = 15'$.

range. The table also gives a value for "threshold focal length" for each i.e., the focal length in mm to produce an image of about 0.1 mm, but this is only a ballpark value, as planets' apparent diameters vary even more dramatically than their magnitudes.

	Magnitude	fx'	f
Mercury	−1.1 to +0.8	8 to 10	2,000
Venus	−4.4 to −3.0	3 to 5	500
Mars	−2.3 to +1.5	6 to 11	1,000
Jupiter	−2.5 to −1.3	6 to 7	500
Saturn	−0.3 to +1.4	9 to 11	1,000
Uranus	+5.7 to +6.1	17	6,000
Neptune	7.7	19	8,000
Pluto	14	28	200,000!

Table 5-3. fx' exposure values for planets photographed with lenses too short to resolve them as extended objects.

That last figure in the table might seem to contradict earlier statements. In particular, in Chapter 3, we claimed that "Pluto and Betelgeuse . . . can be photographed with perceptible disks by the largest observatory instruments," but indicated that the 5-meter Hale telescope on Palomar Mountain offers a focal length maximum of about 150,000 mm. The seeming contradiction has a twofold explanation. In the first place, the professional astronomer's limitations are not quite the same as the amateur's: the pro's armamentarium includes, for example, things with imposing names like "speckle interferometry" and "electronic image enhancement." The professional works to get every possible advantage out of every last millimeter of aperture and focal length.

In the second place, looking at 200,000 versus 150,000 and saying that the 50,000 difference makes them very unequal would be a case of numerical naivete. With the precision (or lack or it) in the above data, it is more appropriate to observe that the two aforementioned numbers, being in the ratio of only 4 to 3, are virtually equal. It is certainly the case that subjects like Pluto and Betelgeuse do tax a professional's capabilities, but they are nonetheless resolved by the combination of magnificent optics augmented by sophisticated technology.

BRIGHTER CONJUNCTIONS AND OCCULTATIONS

Two objects are said to be in conjunction when they appear to pass each other in the sky. That's not a precise definition, but it's good enough for our purposes. Obviously, at least one object must be a member of the solar system, as stars remain relatively fixed. Thus, we can have conjunctions of moon and planets, planets only, planets and stars, and so on. Conjunctions are interesting to us because they provide subject matter for striking photographs, especially when two or more of the night sky's brighter objects are involved. Even more striking is the extreme case, called an occultation, when the nearer object passes between us and the farther one, temporarily blocking it from view. Few specifics can be given for this kind of photography. Obviously, the details for any given event will depend on the objects taking part. Fortunately, various periodicals, like the monthly *Sky and Telescope* and *Astronomy*, and the annual *Observer's Handbook* of the Royal Astronomical Society of Canada, give good coverage to impending events, with adequate notice for advance planning.

The reason for the "brighter" in this section's heading is simply that our basic equipment, the stationary tripod, puts fainter quarry beyond our reach. Here, as in several other situations discussed in this chapter, that old 700/f rule puts an upper limit on our light grasp.

A challenging aspect of this subject is that we very often find ourselves dealing with both point source and extended object
simultaneously—depending on what is "starring" in the event—and trying to work out an exposure suitable to both. In Chapter 7, we will show
how to handle this problem.

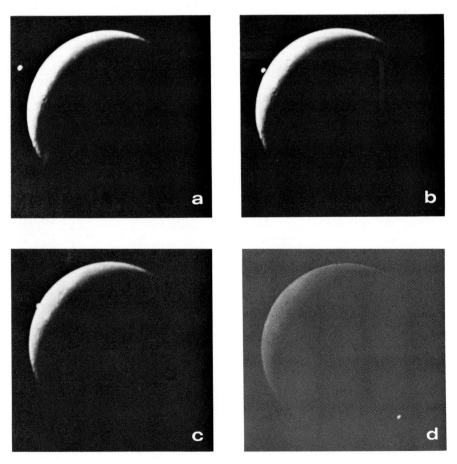

Photograph 5-5. Occultation of Venus as seen from New York City the
morning of December 26, 1978. Taken on Kodachrome 64, using a 750
mm $f/6$ Schmidt-Cassegrain telephoto, on a fixed tripod. Frames a, b, c:
1/8 sec for a total fx = 15. Frame d (about 15 minutes after sunrise): 1/30
sec for a total fx = 13.

Photograph 5-6. Jupiter, Saturn, Venus, Spica, and Mercury (top to
bottom) lined up in the pre-dawn New York City sky at about 5:50 on
November 17, 1980. Taken on Kodachrome 64, 35 mm $f/2.5$, 15 sec, or
$525/f$, using a fixed tripod. Total fx' = 9'.

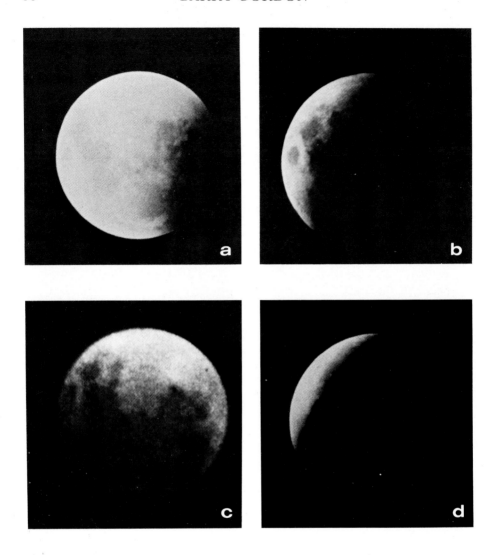

Photograph 5-7. An unusually dark lunar eclipse as seen from New York City, early morning July 6, 1982. Taken on Ektachrome 400, "push" processed to an effective ISO 800. Fixed tripod. Frames a, b, d: 600 mm $f/8$ (via 2x focal extender on 300 mm $f/4$ lens), 1/125 sec, for a total fx = 15. Frame c: 300 mm $f/4$, 4 sec for a total fx = 26; this is $1200/f$—a deliberate violation of the $700/f$ limitation—on the theory that a "soft" image might be better than no image at all.

LUNAR ECLIPSES

A lunar eclipse is the passage of the moon through the earth's shadow. This shadow has an outer part, the penumbra, in which some but not all direct sunlight is blocked, and an inner part, the umbra, in which all direct sunlight is blocked. The umbra, however, is not totally dark, which is what allows us to see the eclipsed moon. Sunlight passing through the earth's atmosphere is bent so as to provide faint illumination even at the umbra's center. This sunlight is not merely bent, it is stripped of its blue and green components. As a result, the eclipsed moon is illuminated by a very pale coppery light; this is what makes the event so spectacular. A total lunar eclipse, one in which the moon totally enters the umbra, can take a matter of hours. There is no need to hurry in photographing it. You should plan to use plenty of film, do a fair amount of bracketing, and try to catch all phases of the event. The opportunity comes all too rarely.

Working with a stationary tripod, you will probably want a fairly fast color film for the umbral phase, to cope with the low light levels and $700/f$ limitation; an ISO 400 color slide film is a good bet. Focal length should be sizeable to get a detailed lunar image, again within the $700/f$ limitation for the very dim mid-eclipse. The exposure recommendations can be found in Table 5-4.

	fx
Uneclipsed Full Moon	10 to 12
Moon Deep in Penumbra	12
Moon Partially Within	
Penumbra	15
Umbra	20
Moon Just Wholly Within Umbra	22
Moon Centrally Within Umbra	23 to 24

Table 5-4. fx exposure values for various phases of a lunar eclipse.

You can't expect the moon to remain a sunlit subject when we're blocking the sun!

Eclipses are particularly well-suited to the "double table" (Table 5-5). As it suggests right in the middle, you enter the table at the bottom on the line that gives the film speed you plan to use. Then, staying on that line, you cross to the column giving your aperture. That column, in the upper part of the table, gives recommended exposure durations, in seconds, for various aspects of the eclipse for that combination of film and lens speeds.

	Exposure Duration in Seconds									
Full (Un-Eclipsed) Moon	1/250	1/500	1/1000	1/2000	1/4000					
Moon Deep Within Penumbra	1/125	1/250	1/500	1/1000	1/2000	1/4000				
Part of Moon in Penumbra	1/15	1/30	1/60	1/125	1/250	1/500	1/1000	1/2000	1/4000	
Part of Moon in Umbra	2	1	1/2	1/4	1/8	1/15	1/30	1/60	1/125	1/250
Moon Just Wholly Within Umbra	8	4	2	1	1/2	1/4	1/8	1/15	1/30	1/60
Moon Centrally Within Umbra	30	15	8	4	2	1	1/2	1/4	1/8	1/15
Use the column in which Lens Aperture and Film Speed occur in the same row below										
ISO 3200-4000-5000	f/64	f/32	f/22	f/16	f/11	f/8	f/5.6	f/4	f/2.8	f/2
ISO 1600-2000-2500	f/32	f/22	f/16	f/11	f/8	f/5.6	f/4	f/2.8	f/2	
ISO 800-1000-1250	f/22	f/16	f/11	f/8	f/5.6	f/4	f/2.8	f/2		
ISO 400-500-640	f/16	f/11	f/8	f/5.6	f/4	f/2.8	f/2			
ISO 200-250-320	f/11	f/8	f/5.6	f/4	f/2.8	f/2				
ISO 100-125-160	f/8	f/5.6	f/4	f/2.8	f/2					
ISO 50-64-80	f/5.6	f/4	f/2.8	f/2						
ISO 25-32-40	f/4	f/2.8	f/2							

Table 5-5. Lunar Eclipse Exposure Table

For example, using ISO 400 film, we would enter the table at the fifth row up from the bottom — ISO 400·500·640. Then, assuming an *f*/4 lens, we would cross over to the fifth column to find *f*/4. Going straight up, we see that the recommended exposure for, say, the moon centrally within the umbra is 2 seconds.

THE SUN

The sun, with its ever-changing patterns of sunspots, is a unique subject. Rather than a shortage of light, we have here such a surplus that it constitutes a real danger. Furthermore, some of the "light" that presents the greatest danger is invisible, the infrared and ultraviolet radiations that flank the visible spectrum. It is impossible to tell merely by looking whether a given attenuator is safe; many materials look very dark because they block most of the visible spectrum, but they may still transmit infrared and/or ultraviolet in retina-destroying quantities. In particular, gela-

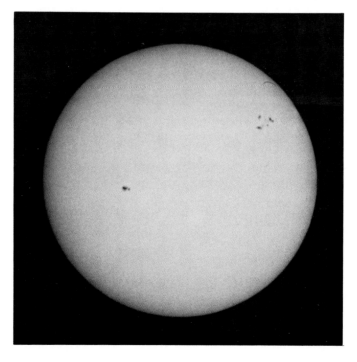

Photograph 5-8. The sun, photographed August 1980 in New York City. Taken on Kodachrome 64, 1500 mm *f*/12 (via 2x focal extender on a 750 mm *f*/6 Schmidt-Cassegrain telephoto), 1/60 sec, with reflective neutral density attenuator of D = 5.3 (or *fx* = −18). Total *fx* = −8.

tin attenuators commonly used in photography, including Kodak's Wratten neutral density #96, *ARE NOT SAFE*. That includes looking through the viewfinder of a camera with such an attenuator: *NOT SAFE*.

What is safe is #14 welder's glass, which has a density of approximately 5.5. Also safe are attenuators that have thin metal coatings evaporated onto glass or plastic bases. These metal coatings reflect most of the radiation that strikes them, visible and near-visible. They are commonly available in the density range of about 3 to 5. A neutral density attenuator of D = 5 is close to ideal; the combination of sun and D = 5 attenuator calls for an *fx* total of 9. The naked sun calls for an *fx* total approaching –8.

The sun's apparent size is always quite close to 32′. Like the moon, it looks bigger near the horizon; as with the moon, that is purely illusory. We can approximate its image size by *f*/107. Since it is so nearly the same size as the moon, we sometimes get to see solar eclipses.

SOLAR ECLIPSES

Strictly speaking, a solar eclipse is really an occultation of the sun by the moon: the passage of the moon between the earth and the sun. Often when this occurs, the moon doesn't quite cover the entire sun; such eclipses are not total, and are photographed exactly as the uneclipsed sun is photographed. But, when conditions are just right, we have a total solar eclipse — one of nature's truly spectacular events.

We hear a great deal about the danger posed by solar eclipses. Yes, it is true, they are dangerous to look at — as dangerous as looking at the sun. During an eclipse, nothing new is added to the sun's radiation; it remains precisely as it always has been. The cause for concern is simply this: normally, attempting to look at the sun is painful because of its extreme brightness; during an eclipse, when the sun is mostly but not completely obscured, it becomes possible to look at the still-exposed "crescent sun" with no discomfort at all. Therein lies the hazard: discomfort or no, however thin the exposed crescent, looking at any part of the solar surface can do serious and irrevocable damage to the eye. During totality, however, with the solar surface completely covered, all danger is gone; the view is totally safe and totally spectacular.

In a way, solar eclipses are easy to photograph. Even during totality, the real show, light levels are generous by astrophotographic standards. Even image size is no great problem, because our subject is not just the 32′ sun itself, but the space surrounding the sun, typically running to 2° or even 3° in apparent diameter. In a way, this subject is virtually tailor-made for our basic equipment. In another way, solar eclipses are among our most challenging subjects. First, we must often travel great distances to locate ourselves within the relatively narrow "path of totality" in order

Photograph 5-9. Total solar eclipse over Ankola, India, on February 16, 1980. Taken on Kodachrome 64, 750 mm *f*/6 Schmidt-Cassegrain telephoto, on a fixed tripod, 1 sec, or 750/*f*, for a total *fx* = 18.

to see the total phase. Once there, and with the weather cooperating, there is an almost overwhelming tendency to stand transfixed, gaping in wonder at the indescribable spectacle. Finally, in contrast to most astrophotography, things happen fast and there is never enough time to allow any margin for error. Opportunities to practice are few and far between.

There are three phases to a total solar eclipse: the first partial phase, during which the moon is covering the sun; the total phase, during which the sun is totally covered; and the second partial phase, during which the moon is uncovering the sun. These phases, and the four "contacts" that are their "boundaries," are summarized in Table 5-6. The two partial phases can be photographed exactly as the uneclipsed sun. They are hazardous to look at without proper eye protection, rather dull, and take about an hour each.

One interesting aspect, however, is the shadows cast by foliage, straw hat brims, and other "light porous" matter; these tend to form images of the partially eclipsed sun and can be striking. The total phase is the main event, which can run from a few seconds to about seven minutes.

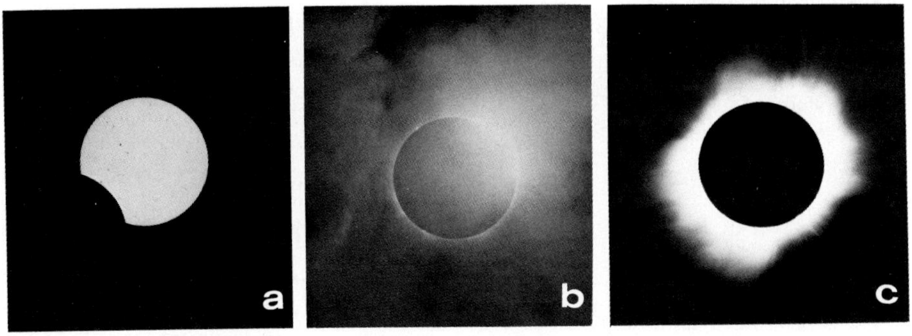

Photograph 5-10. Total solar eclipse over Bratsk, Siberia, on July 31, 1981 (July being just about the best month to be in Siberia). Taken on Kodachrome 64, 600 mm *f*/11 (via 2x focal extender on 300 mm *f*/5.6 lens), on a fixed tripod. Frame a: 1/500 sec using a reflective neutral density attenuator of D = 4.7 (or *fx* = −16) for a total *fx* = −8. Frame b: 1/30 sec for a total *fx* = 12. Frame c: 1 sec, or 600/*f*, for a total *fx* = 17.

SPACE →

TIME ↓

	The Sun	The Sky	The Ground
1st Contact	'Official' Start — nothing to see		
P	First 'Notch'		
a			
r	Growing Coverage	Decreasing Light	Shadow Images
t			
i	Last Crescent		Shadow Bands
a	Baily's Beads	Approaching Shadow	
l	Diamond Ring		
2nd Contact			
Total	Prominences, Corona	Planets, Stars, Horizon Color	
3rd Contact			
P	Diamond Ring		
a	Baily's Beads	Retreating Shadow	
r	First Crescent		Shadow Bands
t			
i	Shrinking Coverage	Increasing Light	Shadow Images
a			
l	Last 'Notch'		
4th Contact	'Official' End — nothing to see		

Table 5-6. Anatomy of a Total Solar Eclipse

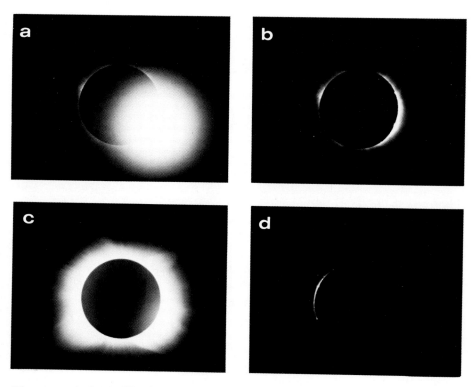

Photograph 5-11. Total solar eclipse over Tuban, Java on June 11, 1983. Taken on Kodachrome 64, 600 mm f/8 (via 2x focal extender on 300 mm f/4 lens), on a fixed tripod. Frames a, b, d: 1/125 sec for a total fx = 11. Frame c: 1/2 sec for a total fx = 17.

As totality approaches, with just a thin sliver of sun still uncovered, there may be a phenomenon called shadow bands — alternating light and dark bands moving rapidly across the ground and other still-sunlit surfaces — but one never knows whether this will occur. Even if it does, it is extremely difficult to photograph. If you can't resist the challenge, your best bet is to find a smooth, light-colored, sunlit surface for your target. If none is available, try spreading a white sheet on the ground. Use a high speed (ISO 1000 or faster) high-contrast black-and-white film, and a high shutter speed (1/500 sec or faster). This is one situation where your light meter (built-in or separate) can be very handy; try exposures of one stop more (wider open) than it indicates, or else go with the given fx value. Either way, bracket if at all possible. Despite all preparations, be prepared to wind up with no photographs at all; shadow bands are rarely seen, much less photographed.

For most of us, the main event begins with the moon's shadow approaching at awesome speed and the breaking up of the solar sliver by the mountains on the moon's edge (Baily's beads). These are quickly snuffed out, one by one, until only one or two small specks of sunlight shine through a lunar valley. These brilliant specks of sunlight are now joined by the first appearance of the relatively bright inner corona, creating the effect known as the diamond ring. Then the last sunlight is obscured, and the full corona — the sun's atmosphere — shines forth in all its glory. Total elapsed time for the events described in this paragraph, perhaps 2 or 3 seconds.

During the few minutes of totality, the corona can be seen and photographed out to perhaps 1° or so from the moon's edge. There are usually planets, almost always including Venus, and a few of the brighter stars to be seen. Down toward the horizon, there will be sky colors suggestive of a 360° sunset. All too soon the first dazzling gleam of sunlight marks the end of totality with the second diamond ring. This is immediately followed by Baily's beads growing into the thin sliver that marks the beginning of the second partial phase. Again there may be shadow bands.

If you are lucky enough to witness a total solar eclipse, plan well ahead what you want to do, and don't be overly ambitious. A successfully executed modest program will get far better results than a messed-up ambitious one. Be sure that the program includes adequate time for plain old-fashioned gawking. If at all possible, get detailed advice from people who have done it. Finally, practice your program — timed — until it becomes all but automatic. Totality is no time for problem solving. Table 5-7 gives exposure recommendations for the various aspects of a total solar eclipse.

	_fx_
Baily's Beads	9
Diamond Ring	11
Prominences	11
Inner Corona	14
Outer Corona	17
Landscape, Horizon	18
Shadow Bands	18

Table 5-7. _fx_ exposure values for various aspects of a solar eclipse.

Here again, the "double table" approach may be preferred (see Table 5-8). As before, enter at film speed and cross over to aperture. Then go straight up to the recommended shutter speeds for the various phenomena of the eclipse.

Good Luck!

Partial Phases . . .											
. . . with ND 4	1/1000	1/2000	1/4000								
. . . with ND 5	1/125	1/250	1/500	1/1000	1/2000	1/4000					
. . . with ND 6	1/15	1/30	1/60	1/125	1/250	1/500	1/1000	1/2000	1/4000		
Shadow Bands	4	2	1	1/2	1/4	1/8	1/15	1/30	1/60		
Baily's Beads	1/125	1/250	1/500	1/1000	1/2000	1/4000					
Diamond Ring	1/30	1/60	1/125	1/250	1/500	1/1000	1/2000	1/4000			
Prominences	1/30	1/60	1/125	1/250	1/500	1/1000	1/2000	1/4000			
Inner Corona	1/4	1/8	1/15	1/30	1/60	1/125	1/250	1/500	1/1000	1/2000	1/4000
Outer Corona	2	1	1/2	1/4	1/8	1/15	1/30	1/60	1/125	1/250	1/500
Landscape — Horizon	4	2	1	1/2	1/4	1/8	1/15	1/30	1/60	1/125	1/250

Use the column in which Lens Aperture and Film Speed occur in the same row below

ISO 3200-4000-5000			f/64	f/32	f/22	f/16	f/11	f/8	f/5.6	f/4	f/2.8
ISO 1600-2000-2500		f/64	f/32	f/22	f/16	f/11	f/8	f/5.6	f/4	f/2.8	
ISO 800-1000-1250	f/64	f/32	f/22	f/16	f/11	f/8	f/5.6	f/4	f/2.8		
ISO 400-500-640	f/64	f/32	f/22	f/16	f/11	f/8	f/5.6	f/4	f/2.8		
ISO 200-250-320	f/32	f/22	f/16	f/11	f/8	f/5.6	f/4	f/2.8			
ISO 100-125-160	f/22	f/16	f/11	f/8	f/5.6	f/4	f/2.8				
ISO 50-64-80	f/16	f/11	f/8	f/5.6	f/4	f/2.8					
ISO 25-32-40	f/11	f/8	f/5.6	f/4	f/2.8						

Table 5-8. Solar Eclipse Exposure Table

TRANSITS

Far rarer than eclipses are transits of Mercury and Venus; the passage of one of these planets between the earth and the sun. Use the same approach as when photographing the sun; the difference is that tripod rigidity is put to the ultimate test. Note that when vibration smears the light from a star's bright image into the sky's dark image, you get a smeared star image. However, when vibration smears light from the sun's bright image into a transiting planet's dark image, you get no planet image at all.

THE BRIGHTEST COMETS

Very, very few comets condescend to be photographed by anything so prosaic as a fixed camera on an ordinary tripod. The overwhelming majority are far too faint. But every once in a great while, a naked-eye comet does appear. If it is truly visible to the naked eye, it is probably photographable with our basic equipment. Unfortunately, variations are so great as to preclude any real guidance. Since such comets are so very rare, you will not likely go broke photographing them, so the best thing to do — keeping the $700/f$ rule in mind — is to expose film as though it were going out of style, trying every focal length and exposure duration that might work. A few good photographs of a spectacular comet can be well worth the expenditure.

Photograph 5-12. Comet West taken from Central Virginia March 5, 1976. Tri-X ("push" processed to an effective ISO 1200), 135 mm $f/2.8$, 4.5 minutes, using an equatorial mount running at sidereal rate. Total fx = 33. By Richard Hull.

Photograph 5-13. Comet Kohouteck taken from Central Virginia January 23, 1974. Tri-X ("push processed to an effective ISO 1200), 135 mm *f/* 2.8, 10 minutes, using an equatorial mount, running at sidereal rate. Total *fx* = 34. By Richard Hull.

CHAPTER 5 EXERCISES

If you've not yet started actually taking photographs, then it's high time.

1. Photograph the moon. Use the data in this chapter as a starting point, then try exposures on either side, i.e., bracket the recommended exposure. Try predicting the approximate image size you will get from the lens(es) you use. Keep records.

2. Photograph a star field near the equator. Using a stationary camera (a camera mounted on a solid tripod), try exposures of 500/*f*, 700/*f*, and 1000/*f* seconds, where *f* is your lens focal length in millimeters. Based on lens focal length and (absolute) aperture, try estimating the area of sky and the dimmest magnitude stars you will record. Using your results, decide what your personal criterion is for exposure duration limitation to produce essentially trail-free images. Keep records.

3. Photograph the "oversized" full moon near the horizon; photograph it again later the same night when it is higher in the sky. Use the same lens. Compare the size(s) of the resulting images. Keep records.

    Hint: A reasonably good way to measure image sizes it to project them and use proportions. Actual image is to actual frame width as projected image is to projected frame width, or

$$\frac{\text{Actual Image}}{\text{Actual Frame}} = \frac{\text{Projected Image}}{\text{Projected Frame}}$$

The last three are usually easier to measure directly than is the first.

Chapter 6
THE EQUATORIAL MOUNTING

As discussed in Chapter 3, the first method of "stopping" subject or camera motion is the use of high shutter speeds, i.e., short exposure durations. But there are situations, and for the astrophotographer these are the norm, when low light levels require exposure durations too long to permit sharp imagery. In these cases, the terraphotographer simply adds light, flash or flood—to raise the available light level. Despite the unbounded optimism and faith of those few trusting souls who are found at each eclipse blithely firing their flash units skyward, this is not a real solution in astrophotography.

WHAT AND WHY

Fortunately, there is a technique that will work for us as well as for our earth-only colleagues: "panning." The term has nothing to do with critical comments; it refers to the horizontal motion of a standard tripod head as in tilting and panning. Thus, with tripod or without, panning refers to the rotation of a camera pointed at a moving subject. In terraphotography, this motion is usually horizontal, as most subjects move in horizontal paths, though exceptions, such as divers, do exist. If, for example, you want to photograph a race on a circular track, you would probably stand off to one side and swing your camera in the horizontal plane to keep it on your subject(s), assuring maximum image sharpness for whatever shutter speed. Ideally, if you had special privilege, you might arrange to position yourself right at the center of the circular track. This would assure profile views at all times, eliminate focus changes by keeping subject distance fixed, and simplify panning by providing essentially constant subject motion. Under these conditions, you might be well-advised to obtain a good solid tripod, level it carefully to maintain panning motion in the same plane as the race track, lock its tilt adjustment securely on the track, and pan with the tripod to assure the sharpest possible pictures. In fact, if your racers are sufficiently uniform in their pace, a clockwork drive mechanism might be added to maximize panning smoothness.

This conventional tripod is known to the astrophotographer as an alt-azimuth mounting, from the words altitude and azimuth (because the

73

tilt and pan motions of the conventional tripod control pointing in altitude and azimuth, respectively). Note that the astronomer does not use
the term altitude in quite the same way as a pilot. The latter uses altitude
to refer to a linear measure (e.g., in meters) giving a height above some
reference level (e.g., sea level). Thus, a plane might fly at an altitude of
5000 meters. The astronomer uses the term altitude to refer to an angular
measure based on the horizon. He might suggest looking for Venus on a
particular evening at sunset at an altitude of 15° directly to the west.
"Directly to the west" is the designation of azimuth, which is simply
another term for direction. The conventional tripod with its alt-azimuth
orientation can be convenient in many applications, but not all.

Instead of a nice horizontal race track, consider one of those "whirligig" amusement park rides—open cars racing around a circular track
tilted from the horizontal. Our technique must change; since our subject
motion now disregards the horizontal plane, it will behoove us to do
likewise. As noted above, we want our tripod's panning motion to be in
the same plane as our subject's motion. That is crucial. In other words,
the axis of our panning motion must be parallel to the axis of rotation of
our subject. Thus, the common alt-azimuth tripod head is converted into
an uncommon tilted alt-azimuth tripod head, the tilt being aligned precisely with the axis of the subject's motion.

This, of course, is just what we deal with in astrophotography. Our
subjects—sun, moon, stars, planets, everything except meteors—are all
part of a tilted whirligig going around us in planes parallel to the plane of
the earth's equator. Unless we happen to be at one of the earth's poles, that
plane is not parallel to the plane of our horizon. Thus, for our purposes,
we want our tilted alt-azimuth tripod head to pan in a plane parallel to
that of earth's equator, i.e., pan around an axis parallel to the earth's axis.
A mounting with this property is called, for obvious reasons, an equatorial mounting. The original panning motion of such a mounting moves
through right ascension or hour angle, analogous to terrestrial longitude.
The axis of this motion points to the celestial poles and is therefore called
the polar axis. The other axis, the original tilt axis, is the declination axis
as it provides motion through declination, analogous to terrestrial latitude. With this arrangement, we now find our subject, lock our declination axis, and pan with the polar axis only. Further, since our subjects are
indeed rather uniform in their pace, we are now able to add a clock drive
to our polar axis, and "pan" automatically for sharp images even over
extended durations. It's basically the old tilted-tripod-on-the-whirligig
ploy.

BALANCE

Balance is a problem when heavy telescopes sit atop tilted-over-to-
one-side mountings. Hanging a significant weight in an eccentric fashion

results in that weight trying to occupy the lowest position available. It is our wish, however, that it will tend to remain exactly where we place it. This requires careful balancing of the equipment about each active axis — that is the polar axis and the declination axis — ignoring any now-locked axes used only for initial polar alignment.

While balancing is often provided for in the telescope's design, it may also require attaching additional weight. Whatever the method, balancing must be done, particularly if the equipment is to be driven mechanically. To make matters worse, the job cannot be done once and for all; it must be redone when a camera is added or removed or even shifted from one location to another. Fortunately, this is a simple chore.

Note that, for the polar axis, perfect balance is *NOT* what we want. If we achieved such balance, then any play in our right ascension tracking mechanism would allow the instrument to wander back and forth between the limits of that play. We are generally better off with slightly imperfect balance, so that our instrument always has slight tendency to fall behind or, even better, to fall forward in its tracking.

ALIGNMENT

As mentioned, the crucial point of the equatorial mount is alignment. Unless the polar axis is truly a polar axis, we cannot expect automatic single-axis tracking, but will be required to make constant and tedious adjustments with our declination axis as well. Even if we were willing and able to make these corrections in declination, we would still suffer a degradation of image quality. The reason is that we can really track only one point in the sky; unless we have a true equatorial mounting, our entire image will rotate about that one tracked point. How much will our image rotate? That is a question of such complexity that no attempt will be made to answer it. Obviously, it depends to some extent on how bad our polar alignment actually is. But it also depends, in a fairly complex way, on the location of our subject relative to the actual aim point of our polar axis on the celestial sphere. Suffice it to say that the amount of rotation is difficult to predict, but it can easily be enough to reduce image quality. For high-quality astrophotography, really good polar alignment is essential.

All equatorial mountings provide some adjustment capability for the proper alignment of the polar axis. On large observatory instruments, these are massive arrangements with a very limited range, to be set once and forgotten (the possible exception being observatories in areas subject to earth tremors). Portable mountings, however, must provide a wide range of adjustments, preferably easily made, to accommodate frequent realignment in a variety of locations. These mountings are in fact one alt-azimuth mounting carrying a second alt-azimuth mounting. The lower mounting adjustments are used to aim the upper "pan" axis at the celestial pole, thus turning the upper mounting into an equatorial one. The trick is to get the aim right.

The first step, of course, is to familiarize yourself with the mounting. If you have built it yourself, then you're obviously quite familiar with it. If it is a purchased unit, read the directions with the mounting readily at hand for reference. (Needless to say, this should be standard operating procedure for any purchased equipment). For example, the "lower alt-azimuth mounting" is a concept, and may be a bit subtle to recognize in the actual hardware. There will indeed be provisions for adjusting the altitude and azimuth, the tilt and direction, of the polar axis, but on a less elaborate unit the provision for direction adjustment may simply be to kick the tripod legs (gently) until the entire mounting is pointed satisfactorily. Similarly, tilt provision on very spartan units consists of propping the tripod leg that points in the north-south direction. Obviously, any method of adjustment will work, but the more sophisticated units generally provide more refined mechanisms. One fact cannot be overemphasized: there is no substitute for knowing your equipment.

The simplest approach starts with aligning the camera or telescope parallel to the polar axis. This alignment can be verified by observing that the camera or telescope will continue pointing in the same direction while turned about the polar axis; if the aim direction changes, parallelism with the polar axis has not been achieved. Once the requisite parallelism has been established, the next step is to adjust the "upper mounting," using the controls on the "lower mounting," so that the camera or telescope is pointed at Polaris. (This, of course, assumes you are in the northern hemisphere.) You now have your equatorial mounting polar aligned. Poorly.

The problem is that Polaris is almost a full 1° away from the north celestial pole, enough to cause a misalignment too great for any but the shortest exposures. This alignment can be improved by noting that the pole is actually offset from Polaris (in a direction and amount that vary over time but that can be obtained from the current literature). This information, or similar data on neighbors of the South Celestial Pole for Southern Hemisphere astrophotographers, can be used to improve polar alignment. At this point, you are ready to align your mounting for serious work.

Find a nice bright star as near as possible to both the celestial equator and the meridian. The celestial equator is the projection of earth's equator on the sky; the meridian is the circle on the sky that runs from due north to due south and includes the celestial pole and the point directly overhead called the zenith. Center your camera/telescope on this star, and follow it by tracking with the polar axis only—with no corrections in declination. If using a camera, use the longest lens you have and a viewing screen that aids in identifying the center of the frame; if using a telescope, try a fairly high-power eyepiece, with cross-hairs if available. After several minutes, note the drift: If the star drifts south, the polar axis points too far east; if the star drifts north, the polar axis points too far west.

Adjust azimuth accordingly, in small steps, until declination drift becomes negligible over many minutes. Then find another easily followed star as near as possible to the celestial equator and as low as possible over the eastern horizon. Track as before without corrections in declination. After several minutes, note the drift: If the star drifts south, the polar axis points too low; if the star drifts north, the polar axis points too high. Adjust altitude accordingly, in small steps, as before.

Note that the instructions above are for northern hemisphere use; for the southern hemisphere, interchange the words north and south throughout.

Now if you've just adjusted the altitude of your polar axis, go back to a "meridian star" and recheck direction; if you need to adjust the direction of your polar axis, go back to a "horizon star" and recheck altitude. Continue doing this until you come to a recheck that indicates that no further correction is necessary; now you have a properly aligned instrument. This procedure will probably take a half-hour when first attempted. With practice, it will become much easier and take perhaps half that time. It is time well spent.

The principle is quite simple: to verify polar alignment, verify the tracking. The reason for choosing stars at the specified locations is that in those regions tracking error is (mostly) attributable to only one of the two possible directions in which the polar axis may be misaligned. This greatly simplifies the correction procedure.

Now that our equatorial mounting is truly worthy of the designation and will faithfully track our subjects (and we have a polar axis clock drive and a power source to run it), we have achieved "automatic panning" which frees us from the $700/f$ limitation on exposure duration. This extends our capabilities, bringing many more subjects within our reach.

Chapter 7
SUBJECTS FOR THE DRIVEN CAMERA

We said that our equatorially mounted, clock-driven camera lets us tackle additional subjects. What it really offers is more subjects of the same kinds. In a sense, our new mounting lets us go back and redo the same things, only more and better. Accordingly, this chapter will be a reprise of Chapter 5; we will refer to that chapter for much data, and concentrate here on those new and different things that result from our expanded capabilities. We must keep things in their proper perspective, however. The driven camera frees us from the tyranny of $700/f$, but it falls far short of solving all our problems. For example, with moderate focal lengths, we move from the one-second neighborhood into exposure durations of many seconds or even a couple minutes. But that's about it; further gains will have to wait. Meanwhile, let's examine what we've "bought."

METEOR TRAILS

We can now, with good timing and a bit of luck, record meteor trails against a backdrop of clearly recognizable constellations instead of star trails. The trick is to choose your time and target area guided by predictions of meteor showers. The actual photography is precisely the same as for starfields and constellations.

STAR FIELDS AND CONSTELLATIONS

You may recall that Chapter 5 discussed "Brighter Star Fields and Constellations" and that extra word, of course, arose from the old limitation imposed by trailing. Aside from that limitation, everything else discussed in that section applies here. In particular, exposure determination is based on the fx' value for the magnitude you want optimally exposed.

Star-field photography is typically short-lens photography. The standard technique is known as piggybacking, which refers to mounting your complete camera/lens combination atop your equatorially mounted telescope, which then becomes the camera's equatorial mounting. Various attachment brackets are available commercially, or readily improvised. For piggyback photography, focal lengths in the 35 mm to 135 mm range are used extensively. Shorter lenses tend to have smaller (absolute) apertures — a 28 mm $f/2.8$ is slower than a 35 mm $f/2.8$ lens where point sources are concerned — so that trying for extremely wide fields brings a significant penalty in light grasp. On the other hand, focal lengths above 200 mm have fields that are too restricted to contain most of the familiar star patterns. You may wish to review Table 3-3 on the coverage of various focal lengths. One implication of this modest focal length range is that the Chapter 5 comment about planets continues to apply: planets can still be considered point sources in this type of astrophotography.

Another implication of modest focal lengths is the fact that minor errors in tracking can be tolerated to a far greater extent than permitted by longer lenses. This means that a moderately good mount and drive will allow perfectly acceptable exposures of fairly long duration. That permits the capture of objects significantly dimmer than heretofore available. Just how dim? A very good question, but one that is difficult to answer.

The crux of the problem is that the sky is not completely dark. Even what we call dark sky conditions still give us some light from the sky background. Stars that are very dim will be overwhelmed by that background and will fail to register visibly on our film. The darker the sky background, the dimmer the stars we can record. Thus, clearly, sky darkness is a significant input in determining the faintest stars we can photograph, i.e., our limiting magnitude. What else goes into that determination? We've said we can reach dimmer objects by longer exposures, but, after a certain time, increased exposure duration merely brightens the sky along with the stars, and we eventually wind up losing more than we gain. Similarly, increased film speed will affect stars and background alike, so we gain nothing. Increased aperture will also collect more light from sky and star alike, so this approach is equally fruitless. At this point, the whole thing seems pretty hopeless. The only way to reach fainter limiting magnitudes is to darken the sky background. Except . . .

If we can't darken the sky itself, once we've chosen the best dark-sky site available to us, maybe we can effect some improvement by darkening the image of the sky. We can do that simply by recalling that the sky is the ultimate extended object, and thus sensitive to changes in focal length. Consequently, if we keep duration and film speed and (absolute) aperture fixed, and if we then increase focal length, we will darken our sky image by spreading the fixed amount of incoming light over a larger area at our image plane. Meanwhile, the images of our point source stars will remain essentially unchanged. The limiting magnitude of the stars we can photo-graph turns out to depend totally on sky darkness and lens focal length.

Incidentally, this same phenomenon is responsible for the fact that binoculars and telescopes show us objects in the not-yet-dark evening sky some time before they are revealed to the naked eye. For example, the so-called "night glass," otherwise known as a 7 × 50 binocular for its 7 × magnification and 50 mm objective lenses, will collect about 51 times as much light as the 7 mm, at best, human eye, having about $50^2/7^2 \approx 51$ times the light-gathering area. For the sky, or any other extended object, the 7 × magnification of the binocular spreads this light over 7 × 7 = 49 times as much retina area vis-à-vis the naked eye, leaving apparent brightness virtually unchanged. For stars and other point-like objects, however, the 7 × magnification has no effect; we still see a point-like object but with 51 times as much light packed into it, the equivalent of a 4 + gain in magnitude.

It is simply the other side of the coin. With our increased photo-graphic focal length, we darken the sky while leaving the stars unaffected; with our night glass binoculars, we brighten the stars while leaving the sky unaffected. The net effect is, of course, exactly the same: more stars in any given sky.

Since we have discussed the sky as an extended object, it should have an fx value. If you are very lucky and photograph under good dark-sky conditions, your sky has an fx value of about 37. For those of us struggling under more typical light-polluted urban conditions, the sky has an fx value of about 35. The Table 7-1 gives rough estimates of limiting magni-tudes for various focal lengths for each kind of sky.

f	Dark	Urban
28	8.0	6.5
50	9.2	7.7
135	11.4	9.9
300	13.1	11.6
750	15.1	13.6
2,000	17.3	15.8
10,000	20.8	19.3

Table 7-1. Estimates of limiting magnitudes for various focal lengths under dark and urban sky conditions.

The formula, obviously an approximation, upon which those values are based is:

$$\text{Limiting Magnitude} \approx \frac{3}{4} \, [\text{sky } fx] + 5 \log f - 27$$

The derivation is given in Appendix E.

A final comment on limiting magnitude. We have said that limiting magnitude depends only on sky darkness and lens focal length; that is indeed the case. We discussed it at length earlier in this section and show it in the mathematics in the appendix. But this only tells us what the limiting magnitude value is, not how we get to it. When we want to know how long we can expose before reaching the limiting magnitude — i.e., what exposure duration will maximize magnitude grasp — then the matter becomes more complex. Obviously, an increase in either our film speed or our aperture will result in reaching our limiting magnitude sooner. How much sooner we still can't say, because another variable has crept into the picture.

Remember that up to now all our exposure discussion has been based on a "central" or optimum exposure. Limiting magnitude is the one exception. If we find that our limiting magnitude is, say, 12, then we do not want to expose so as to center that magnitude in our range of stars recorded. This would make our sky background also central, washing out a great deal of the photograph. That is because the whole basis of the limiting magnitude computation is finding the magnitude that equals the sky in image brightness. In other words, the exposure we want will under-expose our limiting magnitude stars by some amount — probably 3 or 4 fx values, perhaps more. The amount of underexposure depends on a new variable, film latitude, which becomes important only in this situation. But the whole question of limiting magnitude obviously depends on numerous things, including your personal determination of the meaning of "just barely registered on our film." In this unique situation, where we're not aiming target-center in our exposure determination, you must arrive at your own evaluation of your film's latitude, in order to choose the proper exposure bias to apply. Then, and only then, can you use the limiting magnitude and other exposure guidelines to help establish the limiting magnitude exposure duration(s) for your film(s) and lens(es).

CONJUNCTIONS AND OCCULTATIONS

Here again, increased capability brings more problems to be solved. It turns out that this particular branch of astrophotography is perhaps the most complex: not by any means the most difficult, but very likely the most complex. The reason is that these phenomena often involve the moon and/or a star and/or a planet. The first of these acts like an extended object; the second acts like a point source; and the last, as we've seen, can go either way, depending on our lens choice. Ignore the fact that we sometimes deal with asteroids and/or satellites of other planets; for our purposes, these are all point sources add nothing to the complexity of the situation. We merely treat the whole thing as a star field. In rare cases, e.g., the spectacular lunar occultations of Venus, we may be able to deal with both subjects as extended objects; for the moon, this was covered in Chapter 5, and we will get to the planets in Chapter 9. For now, let's look at the more common case, in which our two prime subjects are a point source and an extended object.

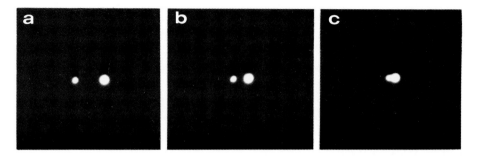

Photograph 7-1. Occultation of (third magnitude) epsilon Geminorum by Mars, as seen from New York City on April 7, 1976. The sequence was taken at about 7:00, 7:30, and 7:48 pm, Eastern Standard Time, on High Speed Ektachrome (ISO 160), 1250 mm $f/10$ Schmidt-Cassegrain telescope, 1 sec, giving total $fx = 18$ and total $fx' = 12'$.

Estimating the appropriate exposure for each subject is not difficult, as long as we remember to use the proper aperture, relative or absolute, in each case. It is really not necessary to use the optimum exposure indicated for each subject; in fact, that will rarely be possible. But it would be nice to have the recommended exposures for the two subjects fairly near each other, as too great a difference will produce a photograph that will either wash out our brighter subject or lose our dimmer one. What can we do to make the recommended exposures come out close to each other? Aren't we simply stuck with the exposures called for by the subjects? Well, yes, we are, and then again, no we aren't. What we can do is remember that, once again, we're juggling a point source against an extended object, and, once again, the key to controlling the situation is focal length. As noted previously, with all else fixed, increased focal length dims the image of the extended object while leaving that of the point source unchanged. Many events involve an overly bright moon interacting with an "underly" bright star, so long lenses are definitely the order of the day—or night.

Again, it is unnecessary, and quite often impossible, to get the two subjects' recommended exposures equal. But, to start, the following table gives approximations to the focal lengths that would be required to balance exposures for point sources versus extended objects.

Point Sources		Extended Objects			Mars, Jupiter,			
		Mercury	Full Moon	Gibbous Moon	Quarter Moon	Crescent Moon	Saturn	
	fx	10	11	12	13	14	15	16
	mag							
Venus	−4					300		
	−3					450		
	⋮					⋮		
Mars, Mercury	0	7,200	5,100	3,600	2,600	1.800	1,300	900
Saturn	1	11,500	8,100	5,700	4,000	2,900	2,000	1,400
	2	18,200	13,000	9,100	6,500	4,600	3,200	2,300
	3	29,000	20,500	14,500	10,200	7,200	5,100	3,600
	4	46,000	32,500	23,000	16,300	11,500	8,000	5,700

Table 7-2. Focal lengths which give the same exposures for point sources and extended objects.

Note that the table is not complete; it needn't be. The only point source brighter than magnitude −2 is Venus (at focal lengths under 500 mm or so), and the only extended object it can approach is the crescent moon. The only other point sources brighter than magnitude 0 are Sirius and Canopus, at magnitudes −2 and −1 respectively, but they are both too far from the zodiac (which is the backdrop for most of our fellow members of the solar system) to get involved in these goings on. Thus, the table as given should cover most situations you will encounter.

For example, suppose you are lucky enough to witness one of the spectacular occultations of Venus by the (crescent) moon. From the standpoint of equalizing exposures for the two, a lens of about 400 mm focal length would be close to ideal. (However, if you want the more generous image size offered by, say, something in the 1000 mm neighborhood, then you'll be dealing with two extended objects and their respective fx values. You'll probably have to settle for a rather overexposed Venus image.) For another example, a conjunction of a gibbous moon with Mars should be handled very neatly, at least as far as exposure is concerned, by a focal length of about 3600 mm. Finally, balancing the exposure for Jupiter, as an extended object, occulting a second magnitude star would call for a focal length on the order of 6500 mm.

The formula for balancing a point source calling for exposure fx' with an extended object calling for exposure fx is:

$$f = 10 \times 2^{\frac{fx' - fx + 20}{2}}$$

The equivalent formula for balancing a point source of magnitude M with an extended object calling for exposure fx is:

$$f \approx 10 \times 2^{\frac{4M - 3fx + 87}{6}}$$

Their derivation is given in Appendix F.

Armed with this last formula, we are now equipped to embark—a side trip, actually—upon a small expedition of scientific inquiry. The diversion is not at all necessary to the pursuit of astrophotography; it is offered merely because some of us find this kind of thing interesting. This particular "thing" occurs if we use the formula to balance a subject with itself. We pose the question: what focal length gives the same exposure for a subject whether it is a point source or an extended object? In other words: what focal length does the formula give us if we use the magnitude and the fx value for one of our switch-hitters?

The question is answered in the following table (Table 7-3), for all the potential switch-hitters in the solar system. In an attempt to keep things reasonably simple, all objects are considered at their roundest. Thus, the moon and all planets as far away as Jupiter are taken at their

	mag	fx	f	A″	s
Sun	−26.7	−8	16	1920	0.15
Moon, full	−12.7	11	14	1865	0.13
Mercury, at superior conjunction	−1.7	10	3,300	4.7	0.13
Venus, at superior conjunction	−3.3	8	3,150	9.9	0.15
Mars, at opposition	−2	13	1,000	17.9	0.09
Mars, at conjunction	+ 2	13	6,450	3.5	0.11
Jupiter, at opposition	−2.7	13	750	46.8	0.17
Jupiter, at conjunction	−1.4	13	1,350	30.5	0.20
Saturn, rings edge-on	+ .67	15.5	1,450	19.4	0.14
Uranus	+ 5.5	16.5	9,650	3.9	0.18
Neptune	+ 7.8	19	11,750	2.3	0.13

Table 7-3. Point Source vs. Extended Object—Threshold Image Size (Based on Focal Length at Which Point Source and Extended Object have Equal Brightness).

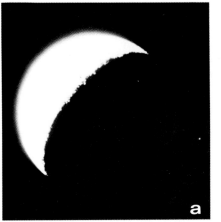

Photograph 7-2. Occultation of (third magnitude) lambda Geminorum, as seen from New York City on May 4, 1976. Taken on High Speed Ektachrome (ISO 160), 625 mm $f/5$ (via 0.5x focal reducer on 1250 mm $f/10$ Schmidt-Cassegrain telescope), 4 sec, giving total $fx = 22$ and total $fx' = 14'$.

full phase; Saturn is taken with rings "edge on" (i.e., not visible from earth). For each object, the table gives approximate magnitude, fx value, and balancing focal length in mm from the above formula. The last two columns show each object's apparent size in arc seconds and the object's resulting image size in mm for the computed balancing focal length shown.

It bears repeating that the input to this table is not, by any means, precise. Nonetheless, one striking fact stands out. With an integrated magnitude range near 35, representing a total brightness range exceeding 60,000,000,000,000,000 to 1, and an fx range near 27, representing a surface brightness range exceeding 100,000,000 to 1, our computed focal length range is very near 835 to 1. But our range of apparent sizes is also very near 835 to 1. The bottom line, or in this case the rightmost column, is an image size range of a mere 2.5 to 1, clustered fairly closely about 0.1 mm! Thus, in the neighborhood of our old familiar 0.1 mm threshold image size value, we get substantially the same exposure recommendation whether we regard our subject as point source or extended object. The remarkable thing is that the 0.1 mm threshold size was arrived at totally without regard to any fx considerations—or any exposure considerations at all, for that matter. From this totally different direction, we arrive at substantially the same threshold criterion. End of diversion.

LUNAR ECLIPSES

You will recall from Chapter 5 that the mid-portion of a lunar eclipse can prove a very challenging subject for the stationary camera, as a result of exposure requirements climbing into the *fx* 23 to 24 range. For the driven camera, all aspects of the lunar eclipse become fair game.

BRIGHTER COMETS

The driven camera promotes us from the very brightest comets of Chapter 5 to merely brighter comets, a rather larger portion of the comet population. Dimmer comets will be considered in Chapter 11.

The comments in Chapter 5 still apply. Not much general guidance can be given. Magnitude estimates are of little value; they are integrated magnitudes of extended objects, so aren't much use in exposure determination. Try anything that seems at all reasonable, except economizing on film.

Chapter 8
OPTICAL SYSTEM

Up to this point, we have been considering only the very simplest optical systems: the image that is formed directly by a camera lens or telescope objective, without auxiliary optics. In this context, the flat secondary of a Newtonian telescope (affecting only image position) can be ignored; and we consider the secondary of a compound Cassegrain or the like to be an integral part of the objective itself. In a sense, we've been capturing the image at its first availability. For that reason, the optics we've been working with so far are called first focus or prime focus systems. Such a system is shown schematically in the top diagram of Figure 8-1. Note that we are departing slightly from our accustomed nomenclature in that we now add a prime ' to the symbols for focal length and focal ratio. Aside from the prime focus designation, that symbolism is appropriate for a second reason: the prime focus system is the basis for several other systems, and we can use our accustomed f and r symbols to refer to the effective focal length and effective focal ratio of the total optical system as a whole.

We will discuss, in some detail, three variations on the prime focus system, as well as some of their many and varied designations. But, before that, let's mention the one system we will not discuss in detail. This is the afocal system, a name which seems to suggest that it yields unfocussed results. Unfortunately, this may often be the case. The technique is to focus a telescope visually, via an eyepiece, on your subject, and then to photograph "into" the eyepiece with a complete camera/lens set at infinity focus. In theory, this can be done with any camera, not necessarily an SLR. The catch comes in focussing the telescope. If you have normal eyesight, and good luck, then you may indeed focus the telescope so that it provides a bundle of parallel light rays, which look to the camera like an infinitely distant subject. But, if the telescope's focus is a bit off, your eyes may very well compensate without your knowledge or control, so that the image will appear to be in focus. Further, if your eyesight is not normal, even if the abnormality is slight, the best telescope focus for you will not produce that parallel ray bundle. In either case, the result is an out-of-focus film image. Of course, using an SLR, you can always check your focus, but in view of the variety of superior alternatives available, the afocal system is not recommended.

The systems we will concern ourselves with all have several characteristics in common, starting with their purpose. That purpose is simply to alter image size, which is accomplished by modifying focal length, as compared to the basic "naked" lens of the prime focus system. They all do this in the same way: by shifting or transferring the image from its original position to a new position. They all perform this image shift by introducing an additional element into the optical path. This element typically consists of several individual pieces of glass, which we refer to simply as a lens. To distinguish it from our original prime lens, we will call the new addition an auxiliary lens. The auxiliary lens may be interposed between the prime lens and its original image plane, or it may be placed "behind" the original image plane, i.e., on the side opposite the prime lens.

The mathematics of these systems is simple, and is the same for all. As we have said, in each case the image is transferred from its original image plane to a new image plane. For lack of any better nomenclature, we will call the distance from the original image plane to the auxiliary lens x, and the distance from the auxiliary lens to the new image plane y. These are indicated in Figure 8-1. The system magnification m is defined as the ratio of the system's new image size to the image size produced by the prime lens alone. Obviously, since image size is (normally) directly proportional to focal length, it must then be true that

$$f = m f' \text{ and thus } r = m r'$$

where f and r are the new system values and f' and r' are the values for the prime lens alone. Absolute aperture, of course, remains totally unaffected. Knowing this, we could take some photographs with the new optical system, of subjects of known sizes, and thus compute our new focal length and our magnification factor. However, we can also determine m in advance, before putting the system to use, because

$$m = \frac{y}{x}$$

We also have an alternative way to determine m — in terms of our auxiliary lens focal length. In theory, this alternative formula is valid for all the techniques discussed in this chapter, but in practice we tend to use it for Eyepiece Projection (to be discussed shortly) rather than the other techniques. For one thing, we generally know the focal length of an eyepiece more readily than for any other auxiliary lens. Furthermore, some of the quantities we deal with are technically negative, but we can

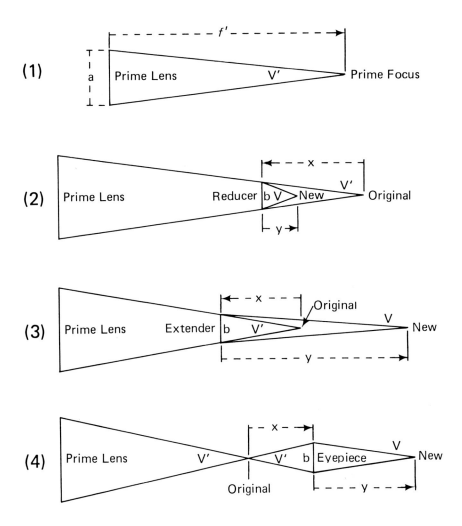

Figure 8-1. Optical Systems—Prime Focus, Focal Reducer, Focal Extender and Eyepiece Projection.

get away with considering them positive if we stick to the above formula for everything but eyepiece projection, and restrict the alternative to eyepiece projection only. The alternative is

$$m = \frac{y}{e} - 1$$

where y is the distance from the eyepiece to the new image plane, and e is the eyepiece focal length.

These are the basic relationships that determine the behavior of all compound optical systems that we will investigate. Their derivations are shown in Appendix G.

FOCAL REDUCERS

While we frequently seek larger images, there are times when we want smaller ones. Generally, this happens when we are photographing very dim extended objects, and want to increase our image brightness by packing the aperture's full light grasp into a smaller image. On such occasions, we use a focal reducer, also called a focal compressor. This is a positive lens interposed between our prime lens and its original image plane, as shown in the second diagram of Figure 8-1. As you can see, the "magnification" in this case is actually a reduction, but we still use the general term — much as the term acceleration includes deceleration as well — and the formulas continue to apply. Focal reducers are available commercially, mounted and ready for use, with various telescopes, usually in the neighborhood of m = 0.5, though other strengths are also available. Not all telescopes are able to use this technique however; the catch is that the telescope must possess sufficient focussing adjustment to allow the film to be moved to the location of the new image plane or vice versa. If your telescope does provide the required flexibility, applying your large telescope aperture at a reduced focal ratio can greatly ease the task of capturing "deep-sky" quarry, as we shall see in Chapter 11.

FOCAL EXTENDERS

In common with its opposite just discussed, the focal extender is also interposed between prime lens and original image plane. In this case, however, we use a negative lens that extends our focal length, as shown in the third diagram of Figure 8-1. Astronomers refer to such a lens as a Barlow Lens and it is also used just ahead of an eyepiece, to amplify visual magnification. Photographers call the same device a tele-extender or a tele-converter. The most common power available is m = 2, with m = 3 and m = 1.4 also fairly common. These seem to be about the optimum.

Focal extenders do not make the demands on telescope focussers that focal reducers do. As Figure 8-1 illustrates, the image shift is in the opposite direction: a focal extender shifts the image plane farther away from the prime lens. Barlow lenses, mounted in barrels compatible with telescope eyepiece holders, make very modest demands on the telescope's focussing capabilities. The purely photographic tele-extenders are mounted in barrels with camera lens mount fitting, to be installed between lens and camera body; they automatically position the film plane correctly, without any outside assist required, so they can be used with any lens or telescope objective.

It's very nice to be able to take, for example, a 500 mm lens and simply convert it into a 1000 mm or even a 1500 mm. But remember you get nothing free. If that 500 mm was $f/8$, then the 1000 mm will be $f/16$ and the 1500 mm will be $f/24$. So those nice big images are going to be nice dim images. Further, you are urged to review the tripod discussion in Chapter 4, and then do a critical examination of your mounting before venturing into "big f country".

PROJECTION SYSTEMS

If you've determined that super-long lens systems are really your dish, then this technique is the ultimate. Since the projection is done by a positive lens, it is sometimes called the positive projection system; since the positive lens used is usually one of your eyepieces, it is usually called the eyepiece projection system. It differs from the other systems in two respects. First, this system allows the prime lens to actually focus its image in the original image plane, unhindered. The eyepiece or other auxiliary lens, in this case behind that image plane, then projects that original image still farther back onto the new image plane, as shown in the bottom diagram of Figure 8-1. Secondly, this system is theoretically capable of yielding m values greater than one or less than one—i.e., true magnification or reduction—depending on auxiliary lens placement. This system does indeed provide the greatest versatility, but even it cannot be pushed to unreasonable lengths without affecting image quality. Powers of m = 4 and up are easily attained with this technique, but we will get back to that.

As suggested by the emphasis on eyepieces, this technique is used with telescopes rather than conventional camera lenses. In theory, adapters could be fashioned to apply positive projection to a telephoto lens, but it just doesn't seem to be done. The one glaring exception occurs in those instances where a telescope manufacturer offers a slightly modified version of a telescope, usually a Cassegrain catadioptric (i.e., mirror-lens combination) as a telephoto lens. Here the line between telescope and telephoto may seem fuzzy, as most so-called mirror-lens telephotos are actually modifications of astronomical telescope designs. Our criterion, and it is a purely pragmatic one, is this: if it accepts only a camera body, then it's a lens; if it accepts a camera body and/or an eyepiece, then it's a

Photograph 8-1. An SLR body coupled for Eyepiece Projection behind a Schmidt-Cassegrain telescope. The eyepiece, of course, is not visible, being inside the tube. It is held (firmly!) in place by the set-screw sticking up on the left.

telescope. So, as stated above, this technique applies to telescopes. The hardware required is simply an adapter, commercially available, which holds a telescope at one end with a camera at the other and an eyepiece somewhere between them.

In view of the extreme versatility of this technique, it is worth a bit more investigation. Let's assume an eyepiece of 25 mm focal length for our auxiliary projection lens. If we position it 50 mm behind the prime lens focal plane — $x = 50$ — then it will form a new image precisely the same distance behind itself — $y = 50$: a perfectly symmetrical situation. Since it is symmetrical, the new image will be exactly the same size as the original, as indicated by $m = 50/50 = 50/25 - 1 = 1$. At the other extreme, moving the eyepiece to 25 mm behind the original image, a move of just 25 mm, we get no second image at all; what we produce, in theory, is an infinitely large image at an infinite distance behind the eyepiece. Thus, simply by choosing an eyepiece location 25 to 50 mm behind the prime lens image plane, we can produce magnifications of any size using just one eyepiece. In theory.

Table 8-1 shows what some of our options look like when e = 25.

x	y	m
50	50	1
45	56.3	1.25
40	66.7	1.67
35	87.5	2.5
30	150	5
25	∞	∞

Table 8-1. Effect of a 25 mm focal length eyepiece used as an auxillary projection lens where x is the distance between the original focal plane and the eyepiece, y is the distance from the eyepiece to the new focal plane and m is magnification.

That's pretty interesting, especially the way magnification starts to take off as we move our lens in closer to that 25 mm barrier. But before getting too excited, take a good look at what else starts taking off: y. That is no mere theoretical value; that is the very real distance that your camera body is hanging in mid-air, behind the projection eyepiece, on some sort of supporting tube. Put another way, that is the length of the lever arm your camera body is using to transmit its own vibration to the rest of the system. So they've got you coming and going: the long lever arm increases the amount of vibration, and then the huge focal length maximizes its effect on the image. Well, monstrous focal length will always make enormous demands on mounting rigidity, but the lever arm problem can be alleviated somewhat by substituting a shorter eyepiece. Let's see what our table looks like (Table 8-2) if we use a 12.5 mm focal length for our projection lens (e = 12.5).

x	y	m
25	25	1
22.5	28.1	1.25
20	33.3	1.67
17.5	43.8	2.5
15	75	5
12.5	∞	∞

Table 8-2. Effect of a 12.5 mm focal length eyepiece used as an auxillary projection lens where x is the distance between the original focal plane and the eyepiece, y is the distance from the eyepiece to the new focal plane and m is the magnification.

Note that x and y scale-down in exact proportion to the focal length of the eyepiece, while the range of magnifications remains the same. Note also that the working locations of an eyepiece of focal length e are all more than e and less than 2e behind the prime lens image plane. Note further the way that small changes in position have large effects on image size and location. Finally, note that shorter eyepieces minimize the lever arm problem for a given magnification.

Unfortunately, we have been looking at the situation completely backwards as far as the real world is concerned. We have been assuming that we can take our eyepiece and position it just about anywhere we want behind the prime image plane. That's not a bad way to try to understand the theoretical behavior of eyepiece projection systems, but, in practice, it doesn't work that way. The eyepiece projection adapters referred to earlier tend to be fairly rigid — thank goodness! — and usually allow for little or no adjustment. Typically, you are dealing with a mechanical geometry that is fixed, and your flexibility lies strictly in your choice of eyepiece. Let's take a look at yet another hypothetical situation, but one that somewhat resembles the real world. Let's assume you have acquired an eyepiece projection adapter that places your film 100 mm behind the eyepiece — y = 100 — ignoring the fact that different eyepieces with their different barrel lengths may change this value slightly. What are your options with your 1000/10 telescope objective and your extensive collection of eyepieces? (y = 100, f' = 1000, r' = 10) (See Table 8-3)

e	m	f	r
40	1.5	1500	15
25	3.0	3000	30
18	4.6	4600	46
12.5	7.0	7000	70
9	10.1	10,100	101
6	15.7	15,700	157
4	24.0	24,000	240

Table 8-3. Effect of a series of eyepieces (e = eyepiece focal length) used as projection lens with a fixed adapter 100 mm behind the eyepiece with a 1000 mm f/10 telescope. m = magnification, f = focal length and r = focal ratio.

Glory be, you say, we've finally solved all the problems. Simply by using short focus eyepieces and sturdy tubes of moderate length, we can get all the magnification we need. That is indeed the case — to a point. The problem is you may lose more than you gain. In the first place, with extreme focal lengths and their associated extreme focal ratios, the nature of light makes itself felt by way of ever-larger diffraction disks. For example, if you apply a factor of m = 10 to a prime lens of $f/10$, you get a system with a relative aperture of $f/100$. Even with perfect optics, points in your subject translate into 0.1 mm dots in your image. If you wished, you could even go to m = 50 giving $f/500$. With this setup, again assuming perfect optics, points are imaged as 0.5 mm dots. What this means is that you are getting ever larger images, but with absolutely no additional detail. This is called empty magnification.

What is worse is that you pay a price for this, though you get nothing in return. For example, in going from m = 10 to m = 20, you are doubling your focal ratio while cutting image brightness to one-fourth. (Of course, we assume your subject is an extended object. Otherwise, why bother with monstrous focal lengths?) How do you compensate? Well, you might switch from ISO 200 film to ISO 800, which might very well require forced development, and accept the penalty in coarser image grain. Or else you can simply expose for, say, eight minutes instead of two, thereby giving your drive mechanism and the atmosphere four times as long to dance the image away from its proper location on the film. Perhaps you'll opt for the compromise — a four-minute exposure on a film with ISO 400 — thus giving equal opportunity to atmosphere, clock drive, and film grain to mess up your results. All this for the satisfaction of being able to say that the subject was photographed at $f/200$!

The simple fact is: pushing magnification to extremes will be costly. Good photographs can be, and have been, taken at $f/100$ and beyond. But, as focal ratios climb into three-digit values, be prepared to work hard and long to maintain quality.

Note that, here again, the limitation is aperture. Not merely in light-gathering power, but also in resolution. Diffraction disk size depends on focal ratio. For a given focal length, the larger aperture will produce the sharper, more detailed image. There is just no getting around it: in resolving power as well as light-gathering power, aperture is king. The more you do in this field, the more you'll come to appreciate that.

But let's not get hung up on pitfalls. Recognizing that we can't push a given aperture beyond its limits, and realizing that our mounting is in for its most severe test yet, let's also recognize that we have taken a quantum leap in our photo capabilities. Now let's see what we can do with them.

Chapter 9
SUBJECTS FOR LONG LENSES

You may recall that Chapter 7 opened with the promise of additional subjects — not additional kinds of subjects, but additional subjects of the same kinds. In this chapter, we can't even promise that; our new capabilities offer not one new subject for our consideration. But they do allow us to redo some former subjects in such a way that they might as well be new subjects. However, at the risk of seeming overly pessimistic, we must recall that we are now two steps removed from easy subjects. From here on, we will work for every decent shot we obtain.

Let's begin by noting that our new quarry, actually our renewed quarry, will call upon all the equipment and skill we have acquired. In particular, we don't set aside our driven equatorial mounting when we move into ultra-long-focus optics; on the contrary, successful use of such optics is totally dependent on the best mountings available. That means mountings that are, first and foremost, rigid and absolute death on vibrations. It also means mountings that are well polar-aligned and driven smoothly and accurately; we will depend on their error-free tracking for many moments at a time, with no correction or other intervention possible.

The one piece of equipment we are not dependent on now is the camera shutter. It can be stated categorically that there isn't a camera shutter made that won't degrade image quality in this kind of photography. Remember that large f values enlarge the effects of vibration as they enlarge the subject. Remember, too that with eyepiece projection your camera shutter is sitting way out on the end of a long lever arm. The conclusion is inescapable: for quality results, it's the old hat trick, as described in Chapter 4. To review, after everything is set up, hold a hat or a dark card in front of the lens, being very careful not to touch any part of it. Then, open the shutter and wait several seconds to allow all vibration to die out. At that point, start the actual exposure by removing the card or whatever, again being most careful not to even brush, much less bump, the equipment. After the proper time, replace the card and immediately close the shutter.

The last source of mechanical problems is the various fastenings and connections holding your gear together. Lenses and telescopes should have no loose parts. See that both are securely fastened to their mountings, which also must be in good working order. See that all parts and couplings in your camera/telescope adapter are properly secured.

Your mechanical worries attended to, it's time to attend to the optical ones. You will find yourself coping with very dim images on your viewing/focussing screen. Even with the moon, you'll find the viewing and focussing of an $f/90$ system to be a far cry from the brilliant $f/1.4$ or $f/2$ images that we've grown accustomed to. Here is one place where those SLRs with interchangeable screens and viewfinders really shine. Even if they don't literally shine, they can be a good bit brighter, and this is where every bit helps. If you enjoy that luxury, try using a screen that is plain ground glass without split image or microprisms. If you can, replace the pentaprism; a critical focussing magnifier is ideal, but even a waist-level hood with magnifier will very likely be a significant improvement. In any event, whatever your viewing/focussing facilities are, be prepared to work at focussing more than you are used to doing. Take your time, and do the best job you can. Take advantage of one of the few advantages we astrophotographers have: most of our subjects are extremely patient and perfectly willing to wait.

At this point, you have done all that you can do. The rest is up to the atmosphere, though you're not totally helpless in dealing with it. The first thing you must do is be aware of the differences in the turbulence of the atmosphere, which astronomers call "seeing". When the air is calm, the seeing is good; as turbulence increases, the seeing becomes poorer. Needless to say, you want to do your long-focus photography on nights of good seeing. Those "lovely winter evenings with all those stars twinkling so beautifully" are not for you. What you want calm air and minimum twinkling.

Avoid "heat sinks", areas that absorb the sun's heat all afternoon and give it back to the atmosphere at night. This results in a constant shimmering of the atmosphere until the excess heat has been dissipated. This can take hours. The worst heat sinks are roads and driveways, concrete and asphalt. To be avoided even more than heat sinks are heat sources, like chimneys. None of these should be under or nearly under the line between you and your subject. Ideally, your line of sight should be over grass, woods, or water.

Photograph 9-1. The moon, taken October 1976 in New York City. Taken on High Speed Ektachrome (ISO 160), 6250 mm $f/50$ (via 5x Eyepiece Projection on a driven equatorially mounted 1250 mm f/10 Schmidt-Cassegrain telescope). 1/2 sec. Total fx = 13.

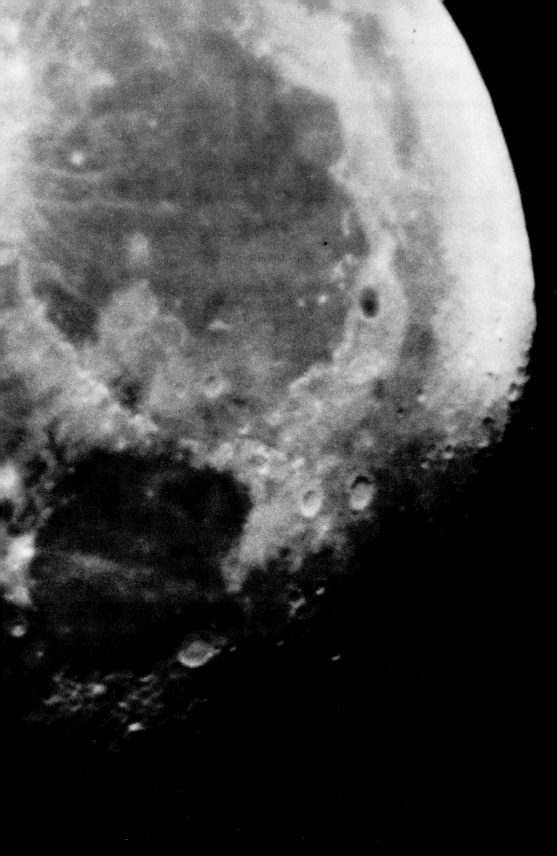

The last thing you can do is use lots of film. Be assured that for every good amateur astrophotograph you have seen, the photographer produced several dozen that are not shown. It is not unusual to use an entire roll of 36 frames in an evening in hopes of getting a couple good ones. The time involved is not prohibitive, as these are short exposures by astrophoto standards, and a couple good shots are easily worth the cost of film and processing.

THE MOON

Back to astrophoto subject #1, we still have good news and bad news. For the good news, subject brightness, the exposure data in Chapter 5, is still the starting point. The exception is the area lighted by earthshine, which, at these focal ratios, you can simply forget. We are now in a position to look at the bad news in an entirely different, albeit dim, light. Going back to our last hypothetical case in the preceding chapter — our 1000 mm $f/10$ telescope objective — even our relatively modest 12.5 mm eyepiece boosts us to an impressive 7000 mm effective focal length with an effective relative aperture of $f/70$. This effective focal length, usually called efl, will give us a lunar image with a diameter on the order of 63 mm. That is very nearly twice the long dimension of our slide, and well over twice the short dimension. More precisely, a 22.9 mm by 34.2 mm rectangle can cover about 25% of a 63 mm circle. So, with one small step or one giant leap, depending upon your viewpoint, we have moved abruptly from photographing the moon to photographing portions of the moon. And the cost? Continuing with our hypothetical situation, and adding a hypothetical ISO 64 film, we find that our exposure durations for a gibbous moon will be about one second. At our 7000 mm efl, this exceeds our old $700/f$ rule by a factor of ten, but should not pose a challenge to any reasonable telescope drive. As a result, we could now start thinking of ourselves as landscape photographers, except that the landscape we're dealing with is some 380,000 km away.

THE SUN

Our new "reach" enables detailed close-ups of sunspots, but all previous warnings and precautions remain. Don't be lulled into a false sense of security just because you've gone from say, $f/10$ to $f/100$, and thus cut your image brightness to 1/100 of its former value. Remember from Chapter 5 that the recommended attenuator for solar work has a density of 5; that density cuts image brightness to 1/100,000 of its original value. The puny effect of your new focal ratio still leaves you dealing with a subject some 1000 times too bright. True, you can now do with less attenuation, e.g., D = 3 rather than D = 5, giving you a nice, bright viewing/focussing image to work with. Go ahead, enjoy. This is absolutely the only long-focus subject that offers that luxury. But don't ever forget you're still dealing with an *fx* = –8 subject. Proper attenuation is still a must. Review Chapter 5's discussion of the sun for details.

The real added problem here is vibration. As noted, there is such a surplus of light that you can choose from a very wide range of exposure durations, simply by choice of attenuator density. The recommendation is to either go for a few seconds via the hat trick, throwing yourself on the mercy of the atmosphere, or else go all the way to 1/1000 of a second, in hopes that the shutter's high speed will freeze the effects of its vibration. Avoided the in-betweens—exposures in the range of 1/15 of a second or so, which require using the shutter and record all of its vibration on film. The only real answer, as in so much of astrophotography, is to experiment.

TRANSITS

So rare as to be hardly worth mentioning, transits (discussed in Chapter 5) are also fine long-lens subjects. In particular, striking photos might show the clean, sharply defined, perfectly round, and perfectly black image of the planet contrasted with the fuzzy, irregular-shaped grayness of a nearby sunspot. Also striking are views of the planet transiting the solar limb—which is astronomerese for crossing the sun's edge.

THE PLANETS

There are two really difficult challenges in astrophotography: extreme dimness and extreme smallness. Extreme dimness we'll defer to Chapter 11. Extreme smallness is the name of the game in planet photography. Aside from stars, almost everything else we deal with is measured in degrees and/or minutes of arc; for planets, the standard unit of measure is the arc second. The trick is to take these minuscule-appearing subjects, looking scarcely larger than stars, and resolve them into real extended objects displaying surface detail. Obviously, when we succeed, we treat them just like any other extended object, using their *fx* values and relative apertures for exposure determination.

A list of planets follows in Table 9-1, giving their *fx* values and apparent sizes as viewed from earth. Also given is a rough idea of the threshold focal lengths in mm required to "convert" them from point sources to extended objects.

	fx	A″	f mm
Mercury	10	5 to 13	2,000
Venus	8	10 to 64	500
Mars @ opposition †	12 to 14	14 to 25	1,000
Jupiter	12 to 14	31 to 50	500
Saturn			
Planet	15 to 16	15 to 21	1,000
Rings	15 to 16	35 to 49	1,000
Uranus	16 to 17	3.7	6,000
Neptune	19	2.5	8,000
Pluto	†	0.1	200,000

† See following discussion of planet.

Table 9-1. *fx* values, apparent size when viewed from earth, and an approximate threshold focal length when a planet moves from a point source to an extended object.

Note that these are merely threshold lengths, i.e., the approximate focal lengths required to produce images marginally distinguishable from point source images. To get anything really worth showing, we must go beyond these. Images of, say, 2 mm across will call for focal lengths on the order of 20 times those shown in the table.

Clearly, some of these planets are well beyond our reach. As a guideline, assume that detailed planet photography is possible for planetary disks of 20″ (arc seconds) or greater apparent diameter. The really skilled worker can produce acceptable results from apparent diameters down to 8″. By contrast, visual observation can be rewarding with planetary disks as small as 6″ in apparent diameter. The difference, once again, is atmosphere. In this situation, the human eye has an advantage over the camera; the former is able to "accumulate" impressions obtained in those brief moments of still air, "suppressing" the intervening disturbed images, while the latter puts everything into the final image indiscriminately. Having viewed the overall situation, let's meet the individual players.

Mercury is not very rewarding. For our purposes, it has no surface features. It is also quite tiny, requiring over 30,000 mm of focal length at its closest. It is always so near the sun that, in a dark sky, it is very low. Its primary interest is that, like our moon, it goes through a progression of phases. But, if phases appeal to you, you'll do much better with Venus.

Venus is the second brightest object in the night sky by magnitude, and first in surface brightness. Venus is second only to the moon as our nearest neighbor in space, giving it the largest apparent maximum diameter of all planets. In appearance it is much like Mercury, showing no markings or features except for its moon-like phases. As it happens, the striking crescent phases always occur when Venus is nearest the earth.

Thus, they can be shown quite well even with relatively modest focal lengths under 10,000 mm. While it has little to show besides its phases, the combination of generous size and spectacular brilliance makes it the ideal first subject for the planet photographer.

Mars is tough. Although its orbit is relatively close to ours, it never appears very large because its actual size is tiny. In addition, despite the relative closeness of orbits, when Mars is on the farther side of that orbit its distance from us is considerable. This size/distance combination can be wicked, resulting in an apparent diameter of about 4 arc seconds. The moral is obvious: we only tackle Mars when it is at or near opposition, i.e., in that part of our sky that is opposite the sun: rising around sunset, at its highest near midnight, and setting around sunrise (as reflected in the table given earlier). Even then, the challenge is formidable, with focal lengths of 20,000 mm and over required to capture the planet's faint markings.

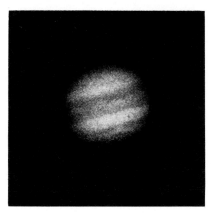

Photograph 9-2. Jupiter, taken September 1974 in Queens, New York on Fujichrome 100 (ISO 100), 12100 mm f/60 (via 6x eyepiece projection on a driven equatorially-mounted 2000 mm f/10 Schmidt-Cassegrain telescope), 3 sec. Total fx is 14. By David Fribourg.

Jupiter is easier. Although considerably farther away than the small inner planets, Jupiter's immense size makes it a comparatively easy subject, as planets go, second only to Venus. Moderately bright, as well as large, Jupiter offers us the endless variety of its belts and zones, as well as the Great Red Spot. These are all within the reach of focal lengths of 10,000 mm or so. Jupiter also offers the endless ballet of its four king-sized satellites. Of these, the brightest and largest is Ganymede, at magnitude 4.6 and apparent diameter of about 1.5 arc second, not that far below extended object candidacy. Referring back to our discussion of conjunctions and occultations in Chapter 7, we see that we could equalize exposures for Jupiter and Ganymede by using focal lengths in the 15,000 mm to 25,000 mm neighborhood. The others of the big four have magnitudes of 5, 5.3, and 5.6, so they theoretically call for focal lengths to about 40,000 mm. Remember that exposures needn't be perfectly balanced; focal lengths of half the theoretical values yield a two-stop difference, which should still produce excellent results. It is also possible to capture one or more satellites' shadows crossing the face of Jupiter, but here again, however, we are likely to need more focal length than the minimum.

Photograph 9-3. Saturn, taken January 1975 in Queens, New York on Kodachrome 64 (ISO 64), 8000 mm f/40 (via 4x eyepiece projection on a driven equatorally-mounted 2000 mm f/10 Schmidt-Cassegrain telescope), 3 sec. Total fx is 15. By David Fribourg.

Saturn, the original ringed planet, is generally acknowledged to be the showpiece of the solar system. Like Jupiter, it has weather bands crossing its face, but these are much fainter and harder to photograph. It also boasts one of the solar system's largest satellites. But, of course, the real attraction of Saturn is its ring system. This rivals Jupiter in apparent size, though the planet itself appears only about half that size. Focal lengths in the neighborhood of 15,000 mm should do the job here.

Uranus, at best, shows a featureless greenish disk of niggardly proportions. It is recommended only for the really determined and/or masochistic. Neptune is very like Uranus, only less so. Pluto, as far as we are concerned, is a magnitude 14 point source and will never be otherwise.

That wraps up our expedition to the realm of the giant focal length.

Chapter 10
GUIDING

It is now time to look into the equipment and techniques required for all those subjects which we have so far ignored: the deep-sky objects — nebulae, star clusters, and galaxies. The last two chapters took us into the realm of the very small; this chapter and the next will deal with the very dim.

How does "very dim" affect photography? For one thing, it might mean the use of very fast films. However, very fast films, or films force-processed to very high speeds, usually carry a penalty in the form of image quality degradation. Trying to push films to the speeds we need is really out of the question.

How about faster lenses? Sounds good, but there's a catch. While we have indeed left the very small, our new subjects are still a long way from being large. Although now the arc second is no longer our unit of measure, we are replacing it only with the arc minute. This means that our new quarry will still call for fair-sized optics. True, we can put aside the 10,000-mm-plus, but we will still be using focal lengths of 1000 to 3000 mm. You will rarely find that focal length coming much faster than *f*/5 or so.

So we are, then, exchanging our super-long lenses for super-long exposure durations. When all is said and done, the very long exposure, perhaps an hour or more, is the key to our new subjects. So we expose longer — so what? So it's a whole new level of complexity and effort.

TRACKING: UNGUIDED VS GUIDED

The motion of a camera or telescope to follow a target is called tracking. Tracking is what a driven camera does. Monitoring that motion, and applying any corrections that are required, is called guiding. Guiding is what *you* do.

The problem is multi-faceted. In the first place, no drive mechanism is perfect. All have their own built-in idiosyncrasies, and that includes the very large observatory drives. Drive errors generally come in two varieties: periodic and erratic. Moreover, many telescope drives run at the solar rate rather than the sidereal rate (a point we will return to later). The result is that no drive tracks properly for extended periods of time without help.

However, even if you could get a perfect drive, guiding would still be necessitated by the second part of the problem: our old nemesis, the atmosphere. It, too, does its dirty work in two ways: systematic and random. The systematic effect is called differential refraction, which refers to the fact that objects in the sky appear slightly higher than their actual locations on the celestial sphere. The obvious exception is that a star at the zenith will appear to be at the zenith, there being no higher place to appear to be. At the other extreme, objects that seem to be right on the horizon would actually be about half a degree below it except for this atmospheric refraction effect. Thus, objects apparently rise more slowly than their theoretical rate, speed up as they approach their maximum altitude in the sky, and then slow down again as they descend. True, the effect is a very small one, but significant where precise tracking is concerned. While systematic image displacement is strictly a function of subject altitude, there is also a similar random effect caused by the constant flow of air masses of different densities across the optical path between us and our subject.

In sum, a driven camera cannot be relied on to track unassisted for extended periods of time. When does guiding become necessary? There is no simple answer. The parameters are: the quality of the drive and its polar alignment, the quality of the seeing or atmospheric steadiness, the focal length of the optics in use, and your own personal image quality criteria. All of these things have been dealt with and will not be rehashed here. From this point, guiding is assumed to be standard operating procedure.

The essence of long-exposure astrophotography is the tracking performed by the clock-driven equatorial mounting that carries your camera. The essence of guided astrophotography is the additional optics, carried by that selfsame equatorial mounting, that let you observe the same sky area that you are photographing. These additional optics may be a whole other telescope, or they may be considerably less; that is a topic unto itself, which we will deal with soon. The entire time that your camera is busy recording the sky with one set of optics, you will be occupied with a second set of optics on the same mounting.

EQUIPMENT

Since guiding begins with monitoring the camera's motion, our first requirement is a means of doing just that. That starts with a so-called guiding eyepiece, an eyepiece with a reticle, typically a cross-hair, at its

Photograph 10-1. Telephoto-equipped SLR mounted piggyback on a driven equatorially mounted Schmidt-Cassegrain telescope. Some noteworthy items shown are: critical focussing direct-viewing magnifier in place of the usual pentaprism atop SLR; reticle illuminating mechanism on eyepiece used for guiding; drive corrector and hand-held control box (sitting on table). By David Healy.

focal plane. Usually, some focussing facility is provided to accommodate other than normal eyesight. Thus, in practice, the reticle is superimposed over the guide object, with both simultaneously in focus. Last, but quite important, there is a source of variable illumination to make the reticle visible, but not dazzling, against the dark sky. A guiding eyepiece with these features makes it possible to recognize even small amounts of tracking drift, an essential for good guiding. Further, common practice calls for rotating the eyepiece in its holder to make the cross-hairs parallel to the celestial coordinate lines of right ascension and declination, so that it very quickly becomes obvious which way you must correct for drift.

Since we've mentioned the other half of guiding—correction—let's examine how that's done before we continue with monitoring and eyepieces.

Corrections in declination are commonly made by a manual declination slow-motion mechanism (a control knob) with a very smooth and gradual effect; electric declination correction motors are also quite popular. The electric control has the advantage of allowing all corrections to be made without touching or jarring the driven equipment, frequently using a single control device such as a computer game joystick. On the other hand, the simple manual control, used with care, is perfectly adequate. The key point is that some provision for declination corrections must be available.

For right ascension corrections, we take advantage of the fact that our drive is a synchronous motor, a motor whose speed varies with the frequency of the alternating current that feeds it, so that we can quickly modify our drive rate by modifying our drive current frequency. This is done with a device called a drive corrector, which is used as the drive power source. The drive corrector is powered in turn by a suitable battery or alternating current source, many models offering the option of running from either. Most of the time, the drive corrector will be "invisible" to the drive, providing alternating current of the standard frequency. However, a group of controls, knobs and/or push-buttons and/or joystick(s), usually mounted on a hand-held box of some sort, can override this and alter the driving current frequency, and thus the speed of the drive motor. This can be done permanently by "biasing" the unit's standard output, or intermittently—or both, with an occasional override of a permanently biased output frequency.

As mentioned earlier, we're a long way from finished with our consideration of the guiding eyepiece. For example, where is this guiding eyepiece? Well, if you are using the technique called piggyback, that is, setting your camera/lens combination atop an equatorially mounted telescope, then you can obviously put the guiding eyepiece into the telescope, which thus becomes guidescope as well as driving platform. But if you're doing prime focus work, using your telescope as the camera's lens, then

Photograph 10-2. An SLR body attached to an off-axis guider behind a Schmidt-Cassegrain telescope. Guiding eyepiece is shown at top, the ideal location. In actual use, it might have to be rotated to a somewhat more awkward position in order to pick up a suitable guide star.

you must do something different. One approach is another telescope, a separate guidescope, riding along with the first on the same driven mounting; this separate guidescope carries the guiding eyepiece, while the other telescope does the photography. The final alternative, a very common one, is the so-called off-axis guider. This is simply a prime-focus coupling device, as described in Chapter 4, with the addition of hardware for deflecting a portion of the telescope's image off to one side and into a guiding eyepiece. The hardware that does the actual deflecting is generally a prism located toward the edge of the image, out of the camera's field of view; thus its name.

Recapping, the new gear we need is: a guiding eyepiece, including a source of illumination; a place to use said eyepiece, either a guidescope or an off-axis guider; a declination slow-motion control; and a drive corrector. Next in order of importance is something comfortable to sit on. In this context, comfortable means capable of providing a relaxed seated posture for an extended period of time. You cannot guide properly from a twisted or cramped position, so your support, as well as the camera's,

Photograph 10-3. "Star's-eye-view" of an SLR attached to an off-axis guider. Guiding eyepiece is out of view at the top. Camera back is removed to show the location and size of the film frame. Prism, centrally located just above film frame, deflects small portion of field to guiding eyepiece.

warrants consideration. You also should have an ordinary kitchen interval timer with an audible signal, so you do not have to split your attention between the guiding eyepiece and a clock. No great precision is required here; your guiding should be dead-on, but your timing can be casual. Another handy item is a flashlight, one with either a red bulb or a red lens. This enables you to see what you're doing without destroying your dark adapted vision. Thus you will be able to doublecheck camera settings, set your interval timer, and make notes. That notebook is still a vital part of your equipment.

One item you may or may not need, depending on local conditions, is a way to combat dewing, moisture condensation on the front surface of your optics. Some astrophotographers buy or make dew caps to prevent condensation. These are simple tubular extensions of the telescope or lens barrel, projecting well forward of the first lens surface. More elaborate dew caps incorporate heating elements, making them even more effective. Alternatively, some astrophotographers use small portable hair dryers to warm their optics, ever so slightly, from time to time. Here, again, a good general recommendation is impossible; you have to determine what best suits your needs.

Finally, you need some protection from either cold or bugs. Sitting quite still for extended periods of time will chill you to a greater extent than you may realize; always dress more warmly than you think necessary. And on those delightful occasions when cold is no problem, insects will be. Insulation or repellent, one of the two is sure to be needed.

PIGGYBACKING

Of the various alternatives for guided astrophotography, the piggyback technique is the least-demanding and therefore, the unqualified recommendation for your initial attempts. As you recall, in this setup a guided telescope carries a complete camera/lens combination piggyback. Typical lens focal lengths are in the 50 mm to 300 mm range, with 135 mm and 200 mm the most popular. An advantage of starting this way is that you are using a fairly powerful telescope to guide a rather modest focal length. In other words, the magnification of your guide mechanism is high by comparison with the magnification of your camera. This gives considerable guiding leverage; tracking errors become obvious in the highly magnified guiding field and can be corrected long before they affect the modestly magnified photo field. Another advantage is the relatively high speed of the camera lenses used, typically in the $f/2.8$ to $f/4$ range, making for comparatively short exposures. In sum, piggyback-guided photography is relatively easy.

PRIME FOCUS: GUIDESCOPE VS OFF-AXIS GUIDER

Once we've mastered the basics of guided astrophotography, we will want to go after the more impressive images available only with four-digit focal lengths. That means photographing through our telescope objective, either prime focus or with a focal reducer, for this is where those auxiliaries really come into their own. Either way, our guiding eyepiece must be relocated, and we must decide between separate guidescope and off-axis guider. It's one more choice you must make.

The separate guidescope is apt to be more expensive. It also introduces one more potential source of guiding error: differential flexure, a fancy way of saying that guidescope and photoscope are not stationary with respect to each other, precluding precise guiding. A rigid guidescope rigidly mounted to a rigid main telescope will add significantly to the load carrying abilities demanded of the mounting. But the separate guidescope is a joy to look through, offering a full-sized, fully illuminated field of view that can make guiding considerably easier. The off-axis guider is easily attached and adds virtually no additional weight or flexure problems. But it can offer a marginal guiding image that often leaves a bit to be desired in terms of brightness and clarity. The separate guidescope generally offers a wider choice of guide objects. The off-axis guider limits you to a star within a relatively restricted sky area relative to your subject. Of course, by the very nature of the beast, the off-axis guider totally precludes guiding on your subject itself, which might be desirable in the case of a rapidly moving comet for example. However, in most cases, the best guide object will be a fairly bright star, sharply focussed.

GUIDING MAGNIFICATION

The last thing to consider before going into the field is guiding magnification. We alluded to this in our discussion of the piggyback technique, and we mentioned guiding leverage. How much leverage do we really need to get good results? Here again, we must say it depends. For example, what do you consider good results? Also, how good are your eyesight and reflexes? Conditionals aside we can give you pretty good guidelines.

Recall that we spoke of guiding magnification as compared to the photographic: the guiding magnification needed is directly dependent on the focal length used. Letting f, as usual, represent our photographic focal length in mm; experience indicates that the very minimum magnification needed for guiding is a power of about $f \div 12.5$. This is precisely what you get with the more or less standard 12.5 mm guiding eyepiece used in an off-axis guider for prime focus photography. (Clearly, whatever the objective focal length, since visual magnification is simply the ratio of objective and eyepiece focal lengths, the combination of prime focus camera and off-axis guider with 12.5 mm eyepiece must give a guiding magnification of $f \div 12.5$) As stated, this is the minimum. Keeping the same 12.5 mm eyepiece and adding a more or less typical focal reducer, with a "magnification" of, say, 0.6, behind the off-axis guider (that is, between the off-axis guider and the camera body), we would cut our effective focal length by a factor of 0.6 and thus raise our guiding magnification to around $f \div 7.5$. This is a significant improvement, and one more reason for the popularity of the focal reducer. However, for really top-quality guiding, many experts advocate guiding magnifications of $f \div 5$ and higher.

Don't be confused by the fact that $f \div 5$ and the like are similar to the notation we use to designate lens apertures. All it means is that, like lens aperture, guiding magnification can be meaningfully expressed as a fraction of focal length.

Table 10-1 gives values for several commonly used telescope objective focal lengths.

Guiding Magnification		Photographic Focal Length			
		f = 900	f = 1200	f = 2000	f = 3000
Minimum:	f ÷ 12.5	72 ×	96 ×	160 ×	240 ×
Better:	f ÷ 7.5	120 ×	160 ×	270 ×	400 ×
Optimum:	f ÷ 5.0	180 ×	240 ×	400 ×	600 ×

Table 10-1. Guiding magnification vs. photographic focal length.

GUIDING PRECISION REQUIRED

But let's not quit here: there is more worth examining. For example, just how precise must your guiding actually be, and how can you tell, during the exposure, whether you are meeting the required guiding precision? There are ways of dealing with these questions.

Our first job is to settle the question of how much precision we will demand in our results. You may recall (from Chapter 3) that we discussed a so-called 700/f rule — and its more critical cousin, the 500/f rule — in connection with minimizing the effects of image trailing with stationary cameras. These exposure duration limitations (for that is what they are) assured image trailing of no more than 0.05 mm and 0.036 mm respectively, for essentially trail-free images. Now we are concerned with producing images that are drift-free, the same problem. Of course, you are free to select any drift tolerance you see fit, but drifting of 0.04 mm or less in the image should prove quite acceptable, at least for your early efforts, so we shall adopt it for our discussion. If you wish to choose a different drift tolerance, substitute your chosen value in the formulas and generate the numbers appropriate to it.

What does 0.04 mm of image drift imply about our guiding? I'm sorry, but once again, it depends. With short lenses, we can tolerate considerable pointing, or tracking, error before our image drifts 0.04 mm; with long lenses, we cause 0.04 mm of image drift with far smaller tracking errors. Our image size formula from Chapter 3 told us image size in terms of subject angular size; this is exactly the same as image drift in terms of angular tracking error! Thus we simply invert the original formula to

$$A \approx 57.3 \frac{s}{f}$$

and then plug in our chosen drift tolerance of 0.04, giving

$$A \approx \frac{2.292}{f} \quad \text{(in degrees)}$$

$$A \approx \frac{137.5}{f} \quad \text{(in arc minutes)}$$

which tells us the maximum allowable tracking error as a function of our photographic focal length, if we wish to maintain our image drift limit of 0.04 mm. Note that, in this context the approximate formula for image size is more than good enough, as we are working typically in minutes (rather than degrees) of arc. Now we know how much pointing "slop" we can tolerate to keep image drift within 0.04 mm. For a few common photographic focal lengths, these tracking tolerances, in arc minutes, are shown in Table 10-2.

Photographic Focal Length (f)	50	135	500	1000	2000	3000
Tracking Tolerances (A)	2.75'	1.02'	0.28'	0.14'	0.07'	0.05'

Table 10-2. Tracking tolerances in arc minutes (A) as a function of photographic focal length (f).

SOLAR VS SIDEREAL DRIVE RATE

At this point, let's digress and resolve the problem of solar versus sidereal rate. As mentioned earlier in this chapter, not all telescope drives run at the sidereal rate (366 revolutions per year); many drive at the solar rate (365 revolutions per year). Is that important? To answer that question, let's assume a camera tracking at precisely the solar rate; what we need to know is the apparent sidereal rate relative to that. Since the stars do one extra circuit per year relative to the sun, our sidereal/solar relative rate is clearly 360 degrees per year, which is about 0.0000114077 degrees per second. Substituting this for A in our original image size equation, we get the trail size or image drift d that results from the sidereal/solar rate difference:

$$d \approx 0.0000114077\, t\, \frac{f}{57.3}$$

$$\approx \frac{ft}{5000000}$$

This can readily be solved for t giving:

$$t \approx \frac{5000000 \ d}{f}$$

and if we continue with our assumption of 0.04 mm as the limit on acceptable image drift, we can substitute this for d to get:

$$t \approx \frac{200000}{f}$$

This tells us approximately how often, in seconds, we must manually correct our solar drive to maintain the desired image quality. Let's see (Table 10-3) what that means for our illustrative set of common photographic focal lengths:

Photographic focal lengths	(f)	50	135	500	1000	2000	3000
Correction frequency in seconds	(t)	4000	1500	400	200	100	67

Table 10-3. Frequency of correction in seconds (t) for common photographic focal lengths (f) when used with a polar aligned telescope drive running at solar rate.

Thus it isn't much of a problem for a "normal 50," as a correction every hour or so is hardly worth mentioning. But, as focal lengths go up, a correction every minute or so, just for sidereal/solar rate adjustment alone, can become a real nuisance. Obviously, except for piggybacking short lenses, a telescope drive that runs at solar rate really needs a drive corrector, and that drive corrector should have a permanent bias control to modify its standard output.

Now that we know just how much guiding precision we need, let's investigate the other side of the question: how do we know when our guiding is maintaining that precision?

DETERMINING HOW WELL YOU ARE GUIDING

The answer to this problem lies within our guiding eyepiece, specifically, right on the reticle. We said that this reticle is typically, and unfortunately, a cross-hair. Ideally, it should be something   more: a cross-hair combined with some sort of area indicator. One such highly recommended is a "double cross-hair," i.e., two perpendicular pairs of lines, forming a neat little square box right in the center of the field. Other types can be used in similar fashion, but our discussion will assume a double cross-hair with central guide box. Thus, our next task will be to deter-mine the size of this box, as the box size measure is central our problem's solution.

A simple way of finding box size is to consult the supplier's specifica-tions. One rather good example quotes line-pair separation as 0.2 mm, which is obviously our box size. Lacking data from the supplier, we can readily determine it by the fact that stars near the celestial equator cross the sky at very close to 15 minutes of arc per minute of time. With our telescope stationary, we can time the "transit" of a near-equatorial star across our guide box and thus quickly determine its apparent angular size against the sky with the particular objective in use. However, knowing the actual reticle size in, say, millimeters will let us determine its apparent angular size when used with any objective, so we call upon our old trail length formula from Chapter 3:

$$d \approx \frac{ft}{13750}$$

where t is our transit time in seconds, f (as usual) is our objective focal length, and trail length d our box size in the same units as f, generally mm. In all likelihood, your box size will turn out to be quite close to 0.2 mm, so we shall assume that value for the rest of our investigation. Once we have the box size of the actual reticle, we can use our inverted image size formula

$$A \approx 57.3\,\frac{s}{f}$$

where s is now our box size and A its apparent angular size on the sky, for any guiding objective of focal length f. Here, too, we can simplify by plugging in our assumed value, in this case our 0.2 mm box size, giving

$$A \approx \frac{11.46}{f} \quad \text{(in degrees)}$$

$$A \approx \frac{687.5}{f} \quad \text{(in arc minutes)}$$

as our apparent box size in angular measure. For a few common guiding focal lengths, Table 10-4 shows the amounts of sky, in arc minutes, that a 0.2 mm box size corresponds to.

Photographic focal length	(f)	900	1200	2000	3000
Box size in arc minutes	(A)	0.76'	0.57'	0.34'	0.23'

Table 10-4. Amount of sky covered, in arc minutes, by a 0.2 mm square reticle box for various photographic focal lengths.

Clearly, a longer-focus guiding objective will magnify the sky area it shows, spreading a smaller amount of sky over our guiding box. It will magnify small amounts of drift at the same time and thus make them much more obvious.

At this point let's digress to take a slightly closer look. To repeat: a longer-focus guiding objective will spread a smaller amount of sky over the guiding box, making small amounts of drift much more obvious. Note that guiding eyepiece focal length has not even been mentioned! This is because it has no effect on what we are considering. Remember that a telescope eyepiece is merely a magnifying glass with which we examine the image produced by the objective; a guiding eyepiece is no different, except that it lets us examine the reticle along with the image. Since changing eyepiece power, i.e., focal length, will affect image and reticle alike, it can have no effect at all on the reticle/image relationship. Thus, a shorter eyepiece might help us see both cross-hair and sky a bit better, but only a longer objective will magnify the sky relative to the cross-hair or guiding box. True, either change will increase guidescope magnification, but via increased objective focal length it is more useful. This is why guiding eyepieces tend to be more or less standardized at about 12.5 mm.

As you may have guessed, what we're really after is a way to monitor our guiding efforts by relating the drift we see in our guidescope to the drift we're producing in our image. In other words, are we all right if our guide star always remains within our guide box? Could we allow wandering over an area two box sizes across? Or must we restrict our guide star to an area one-half box size wide? As if you didn't know, the answer is it depends. But now we know what it depends on, so we are ready to proceed with some definite answers.

As suggested, what we are seeking is the amount of tracking error T that we can accept, as a multiple of our guiding box size, i.e., expressed in guiding box size units. That's it. The formula for this guiding tolerance involves five quantities:

1. guiding tolerance as a multiple of guiding box size: T
2. amount of acceptable image drift: d
3. guiding box size: b
4. guiding focal length: g
5. photographic focal length: f

and is simply

$$T \approx \frac{d\,g}{b\,f}$$

The (trivial) derivation is in Appendix H.

If we continue with our assumed image drift tolerance of 0.04 mm and guiding box size of 0.2 mm, then our formula simplifies to

$$T \approx \frac{0.2\ g}{f}$$

a surprisingly simple expression, involving nothing more than the relationship between our two key focal lengths: photographic and guiding.

This, of course, tells you how precisely you must guide given the focal lengths f and g. For example, whenever guide and photo focal lengths are equal, your guide star must be confined to an area with sides equal to one-fifth of the guide-box size.

Should you be fortunate enough to be able to select from a range of focal lengths, then you might prefer the equivalent form

$$g \approx \frac{b}{d}Tf$$

which, on substituting our familiar 0.2 for b and 0.04 for d, becomes

$$g \approx 5\ Tf$$

This tells you, for example, that you need $g = (5/3)f$ if you want $T = 1/3$—i.e., if you want to let your guide star wander over an area that is one-third box size on each side. If you want things even easier—for example, $T = 1/2$ to let your guide star wander with impunity over an area one-half box size on each side—then you need $g = (5/2)f$.

Is it realistic to imagine that you might actually "be fortunate enough to be able to select from a range of focal lengths"? Quite possibly. Let's consider the many owners of the very popular Schmidt-Cassegrains that have focal lengths of 2000 mm. With this and an off-axis guider and 12.5 mm eyepiece, we get equal guiding and photographic focal lengths—2000 mm—and a guiding magnification of 2000/12.5 = 160x which (as we've already noted) is obviously $f \div 12.5$; these (as we've also noted) are our minimal acceptable values for decent guiding. Let's add a focal reducer, such as the 0.6x example mentioned earlier. This, of course, leaves our 2000 mm guiding focal length, and 160x guiding magnification unchanged, while our photographic focal length is cut to 1200 mm. Now we have our guiding focal length equal to $(5/3)f$ and a guiding magnification of $f \div 7.5$, certainly an improvement. But let's not quit there. Suppose we put a Barlow lens just ahead of our guiding eyepiece, say, for example, one with a magnifying power of 1.5x. Now our photographic focal length remains at 1200 mm while our guiding focal length goes up to 3000 mm, and we have a very desirable guiding focal length equal to 2.5*f* and guiding magnification, now 240x, equal to *f* ÷ 5.

Table 10-5 gives guiding tolerance values T for a variety of focal lengths, guide and photo, continuing with our assumed image drift of 0.04 mm and guide box of 0.2 mm. For each focal length, the table gives the amount of sky corresponding to image drift (for photo) or to guide box (for guide). For example, a 135 mm telephoto lens produces 0.04 mm of image drift from a 1.02 arc minute tracking lapse, but a 900 mm guiding objective would show the guide star crossing the entire 0.2 mm guide box as a result of only a 0.76 arc minute tracking error. The body of the table, the guiding tolerance values we've been after all along, are really the ratios of those amounts of sky. So, now that we finally have our answers, let's spend a few moments looking them over.

Photo Focal Length, in mm [Image Drift of 0.04 mm, arc min]		Guiding Focal Length, in mm [Apparent Size of 0.2 mm Reticle Box, arc min]			
		900 [.76]	1200 [.57]	2000 [.34]	3000 [.23]
50	[2.75]	3.6	4.8	8	12
135	[1.02]	1.3	1.8	3	4
200	[0.69]	0.9	1.2	2	3
300	[0.46]	0.6	0.8	1.3	1.2
400	[0.34]	0.45	0.6	1	1.5
500	[0.28]	0.36	0.48	0.8	1.2
750	[0.18]	0.24	0.32	0.5	0.8
1000	[0.14]	0.18	0.24	0.4	0.6
1500	[0.09]	0.12	0.16	0.27	0.4
2000	[0.07]	0.09	0.12	0.2	0.3
3000	[0.05]	0.06	0.08	0.1	0.2

Table 10-5. Guiding tolerance values (T) for selected guide and photographic focal lengths based on an allowable image drift of 0.04 mm using a reticle guide box of 0.2 mm.

Starting with a nice simple case, guiding a 400 mm lens with a 2000 mm guidescope, we should be in fine shape if we merely confine our guide star to our guiding box; letting it escape runs the risk of image degradation. We have a bit more leeway, an area 1.2 box sizes across, if we guide 200 mm with 1200 mm, or guide 500 mm with 3000 mm. Now that does not imply that these are equally easy to guide; degrading image drift at 500 mm results from far smaller tracking errors than at 200 mm, calling for a delicate hand on the controls; but the correspondingly larger guiding focal length will make those smaller tracking errors equally easy to spot. As you would expect, guiding gets easier with shorter photo focal lengths, or with longer guiding focal lengths. But we also note that the table is divided into three parts. This is simply a reminder that:

- a guiding tolerance of 1/2 box size or better makes guiding somewhat more comfortable;
- a guiding tolerance of 1/3 box size or better continues to be eminently practical;
- a guiding tolerance of 1/5 box size or better remains within the realm of the possible;
- a smaller guiding tolerance is not to be considered by rational people.

Is this purely arbitrary? Not really; it's based on a fair amount of experience. It is also a restatement of something we said earlier. You may verify it by working through the modest algebra or a few numerical examples, but it happens that (for a 12.5 mm guiding eyepiece) a guiding tolerance of 0.2 box size corresponds precisely to a guiding magnification of (photo) $f \div 12.5$, and a tolerance of 0.5 box size is exactly equivalent to (photo) $f \div 5$—just about where all this started.

SUMMARY

In summary what we have here is three sides of the same coin—all right, make it two sides and the edge: three different but absolutely equivalent ways of looking at the guiding problem. Each is nothing more than a ratio, explicit or implied, that is simply a shorthand statement of the relationship between two numbers. Let's take a final look at these three ratios, to pin down the way they relate to each other and to our guiding.

Our (by now) familiar assumptions are that eyepiece focal length e = 12.5 mm; guiding box size b = 0.2 mm; and maximum acceptable image drift d = 0.04 mm. Under these assumptions, the ratios, we need for guiding are shown in Table 10-6. And what can we photograph with all this? Just about anything in the universe!

		Minimum	Better	Optimum
Focal Length to Guiding Magnification:	f/m	12.5	7.5	5.0
Guiding to Photographic Focal Length:	g/f	1.0	1.7	2.5
Guide Star Drift to Guide Box Size:	T/b	0.2	0.3	0.5

Table 10-6. Ratios needed for guiding.

Chapter 11
SUBJECTS FOR THE GUIDED CAMERA

You may recall that Chapter 7 offered no new kinds of subjects, merely new subjects of the same kinds. Chapter 9 did not even offer new subjects, merely new ways of dealing with previous subjects. In this chapter we deal with subject matter that is radically different from anything previous, and which is commonly regarded as the ultimate quarry — the ultimate challenge — of the astrophotographer.

For starters, we will probably have to begin taking limiting magnitude considerations seriously. Granted, we did examine that problem in great detail in Chapter 7, but our hindsight now tells us that, thanks to the inherent limitations of unguided tracking, it was not encountered frequently in that phase of our activity. Now, however, it could be a very frequent problem, especially working piggyback with fast lenses of $f/2.8$ or so, which start recording sky background surprisingly quickly.

The second problem is the demands on our equipment. While vibration is less a problem than it was at the enormous five-digit focal lengths of planetary photography, here at four-digit focal lengths it still cannot be ignored. At the extended exposure durations we will have, our equipment will be changing its position by significant amounts during each exposure. This means that anything that can flex, probably will flex; anything that can shift, probably will shift. However carefully you guide, flexing and/or shifting components will inevitably degrade the quality of your final results.

The last of our major problems is the utter impossibility of providing good hard data on appropriate exposure. For one thing, we are dealing in such fragile images, formed by such minuscule quantities of light, that even small differences in atmosphere and equipment can have major effects. Even worse, we are now well into the dark domain of reciprocity failure, which mocks our pitiful attempts at precision. We will give what guidance we can, but it's purely ballpark estimate all the way.

There are two bits of good news. The first is that the limiting magnitude problem tends to be less serious than theory indicates. What comes to our aid, suprisingly enough, is reciprocity failure. For example, suppose we wish to photograph a really dim subject with an *fx* value of 35. And suppose we have a site with a good dark sky, which we've said has *fx* = 37. That indicates that our sky background, being only about two *f/* stops darker than our subject, should show up only slightly darker on our film, not really dark at all. Reciprocity failure says that, at this already very dim level, cutting the light by any amount at all will have an effect out of proportion to the actual light decrease. So our sky background will record much darker than the small *fx* difference suggests. Note that you're still taking on quite a challenge if you try that *fx* = 35 subject from an urban site with an *fx* = 35 sky. But with any difference at all, reciprocity failure will work in your favor and exaggerate that difference to improve the limiting magnitude over what theory suggests. In this context, it does not matter whether we're dealing with point sources or extended objects. If the sky is at all darker than the subject, then reciprocity failure will help by making that sky appear even darker on film.

The other bit of good news concerns exposure precision. You will recall the bad news: you can't have it. The good news is you don't need it. Going way back to Chapters 2 and 3, we discussed various things that affect image size. In Chapter 5, under solar eclipses, we hinted at one more — exposure. What we said was, try *fx* = 14 for the inner corona and *fx* = 17 for the outer corona. Taken literally, of course, that's just non-sense; the sun doesn't have two coronas. Obviously, it means that the sun's one and only corona varies in brightness, getting dimmer with increasing distance from the central portion. Thus, increased exposure shows a "bigger corona" by recording the dimmer outer portions at the sacrifice of overexposing the brighter inner parts. In other words; there is a whole range of "correct" exposures for the solar corona, depending on the trade-off you want between outer extent and inner detail. The very same thing applies to many of the subjects we are about to tackle. In fact, with many, you'd have to work at it to produce a really "bad" exposure: almost any exposure over a very wide range will produce a good photograph of some portion or aspect of the subject.

Now let's re-introduce the one hold-over subject from earlier chapters, and then meet the brand new members of our celestial cast.

COMETS

The guided camera will put more comets within your grasp than are available to the unguided camera. The comet comments of Chapters 5 and 7 continue to apply. The comments immediately above, on lack of good exposure data, apply with a vengeance. If size and magnitude are known, some very rough estimates may be made by comparison with one or more deep-sky objects, discussed below, with similar attributes.

If you're after a rapidly moving comet, and the option is available to you, guiding on the comet itself—obviously with a separate guidescope—may yield slightly better results than an off-axis guider. But, in any event, any comet that presents itself is a worthwhile subject, whatever guiding method is available. A really sharp image is impossible anyway as the subject itself is fuzzy.

THE MESSIER CATALOG

One of astronomy's great comet hunters—and comet finders—was the eighteenth-century French astronomer, Charles Messier. In his search for new fuzzy objects, he kept coming across old fuzzy objects which were not comets at all. To facilitate recognizing and dismissing these each time they re-appeared, he simply compiled a reference list of them. This list, created strictly as a comet-hunting aid, is known as the Messier Catalog, and is the standard introduction to nebulae, star clusters, and galaxies. It is not a complete introduction to deep sky objects and it was never intended to be, but this compilation of a comet hunter's discards, most of them extended objects, makes a good starting point.

There are about 110 entries in the Messier Catalog, depending upon the version. Each entry is designated by its catalog number preceded by the letter M. Thus, the first entry is M1, the second is M2, and so on. A much larger catalog, intended as a deep-sky reference, is the *New General Catalog of Nebulae and Clusters*. This is referred to as the *NGC*, and that three-letter prefix identifies its entries (The *Index Catalog, IC*, and the *Second Index Catalog*, are also bound with the *NGC*.) With these identifications, you will be able to get further information on objects of interest, including their locations on the celestial sphere. In addition to formal designations, most objects have popular names, firmly established in the literature. We shall give those as well. We will not duplicate the work of a good sky atlas, which you are urged to acquire. Our concern here is with photography. Finding them is up to you; however, most experienced observers recommend, as a starting point, the three volume work *Burnham's Celestial Handbook* by Robert Burnham.

PLANETARY NEBULAE

Photographing planetary nebulae makes a nice transition from planet subjects to deep-sky. As a group, they are the smallest and brightest objects in the Messier Catalog (in surface brightness, not integrated magnitude). Compact, with relatively well defined edges, they get their class name from the fact that they often resemble planets. They are thought to be expanding shells of material ejected from a star's outer layers. The Messier Catalog lists four planetaries with magnitudes ranging from 8 to 11. Table 11-1 gives *fx* values, plus integrated magnitudes and sizes in arc minutes, for the largest and smallest, also the brightest and best-known examples.

	fx	Mag.	A′
M27 – Dumbbell	32	8	8 by 4
M57 – Ring	31	9	1 by 1

Table 11-1. *fx* values, integrated magnitudes, and sizes in arc minutes for the largest and smallest planetary nebulae in the Messier Catalog.

This pair provides another good example of our brightness ambiguity. Although astronomy-brighter by magnitude, the Dumbbell is spread over a much greater area and is therefore photography-dimmer, requiring greater exposure than the Ring. We will run into this situation time and again.

Photograph 11-1. M27, the Dumbbell Nebula (perhaps in honor of those who try photographing it from urban skies). Taken November 1982 in Naco, AZ on gas hypered Kodak TP2415, 2100 mm f/6 (via focal reducer on 3900 mm *f*/11 Schmidt-Cassegrain telescope), 30 min. Assuming ISO 250 for film, total *fx* = 32. By David Healy.

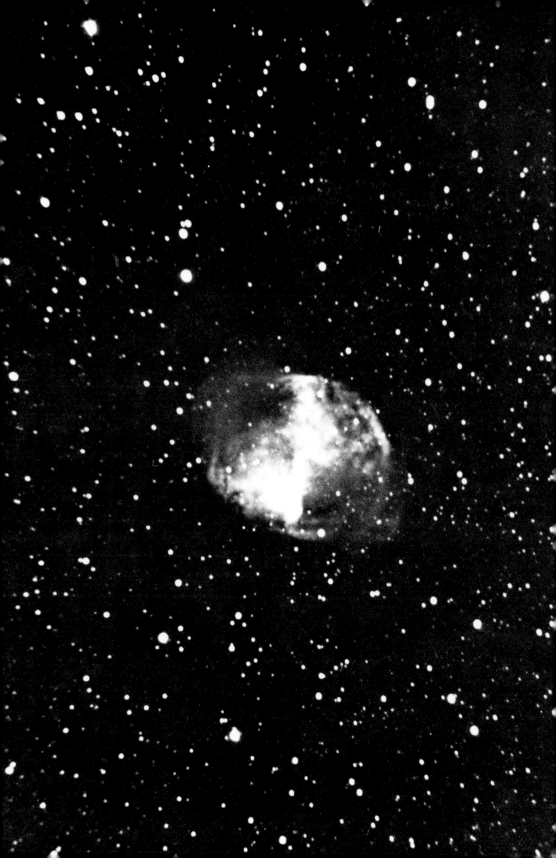

Photograph 11-2. M57, the Ring Nebula. Taken April 1976 in Naco, AZ on GAF500, 2000 mm f/10 Schmidt-Cassegrain telescope, 40 min. Total *fx* = 31. By David Healy.

DIFFUSE NEBULAE

Among the most beautiful of all photographic subjects, the diffuse nebulae, as a class, are the largest objects in apparent size in the Messier Catalog. They are also among the brightest in integrated magnitude, and many display the center-to-edge brightness variation mentioned earlier. Messier lists eight, with a magnitude range of 4 to 9 and sizes of 6' by 4' to an impressive 85' by 60'. Diffuse nebulae are often classified as emission, reflection, or dark, according to whether they are luminous or nonluminous, or simply block the light of brighter areas beyond them. For a dark nebula, of course, our exposure must consider the lighter background against which the subject is silhouetted. The distinction between emission and reflection is very important to astrophysicists, but, as astrophotographers, it's one of the (few) things we needn't worry about. Table 11-2 gives values, as above, for several of the group's showpieces.

	fx	Mag.	A'		
M1 — Crab	31	8	6	by	4
M8 — Lagoon	33	6	60	by	35
M17 — Omega	33 to 34	7	46	by	37
M20 — Trifid	33 to 34	9	29	by	27
M42,43 — Orion					
Center	28				
Overall	32	4	85	by	60
IC 434 — Horsehead	33		60		
NGC 3372 — eta Carinae	33	5	85	by	80
NGC 6960,9295 — Veil	33		120		
NGC 7000 — North America	34	1	120	by	100

Table 11-2. fx values, integrated magnitudes, and sizes in arc minutes for a number of diffuse nebulae.

Diffuse nebulae are associated with star formation and star destruction. Enormous clouds of gaseous matter such as M42, many times the size of our solar system, gradually condense to form new stars. The smaller of these stars eventually burn out, "not with a bang but a whimper". The larger, however, wind up their careers in spectacular fashion as supernovas, exploding vast portions to create supernova remnants such as M1. These remnants continue to expand through the interstellar medium, the Veil being an ancient example. Eventually much of this material finds its way back into enormous clouds of gaseous matter such as M42, many times the size of our solar system, which gradually condense to form new stars.

Since these new stars are formed in groups in relatively small regions of space, they begin as members of open clusters.

Photograph 11-3. M8, the Lagoon Nebula. Taken September 1981 in Naco, AZ on gas hypered Kodak TP2415, 2500 mm $f/7$ (via focal reducer on 3900 mm $f/11$ Schmidt-Cassegrain telescope), 24 min. Assuming ISO 250 for film, total fx = 31. By David Healy.

Photograph 11-4. M16, in Serpens. Taken October 1983 in Naco, AZ on gas hypered TP2415, 2100 mm $f/6$ (via focal reducer on 3900 mm $f/11$ Schmidt-Cassegrain telescope), 40 min. Assuming ISO 250 for film, total fx = 32. By David Healy.

Photograph 11-5. Wide-angle view of the belt and sword area of Orion. Taken February 1977 in Everglades National Park, FL. Taken on Fujichrome F100, piggybacked 85 mm $f/1.8$, 6 min. Total fx = 31.

Photograph 11-6. M42, the Great Nebula in Orion, plus surrounding area. Taken January 1978 in Everglades National Park, FL on Fujichrome F100, 625 mm $f/5$ (via 0.5x focal reducer on 1250 mm $f/10$ Schmidt-Cassegrain telescope), 15 min. Total fx = 30.

Photograph 11-7. M42, the Great Nebula in Orion. Taken April 1976 in Naco, AZ on Fujichrome F100, 1000 mm $f/5$ (via 0.5x focal reducer on 2000 mm f/10 Schmidt-Cassegrain telescope), 30 min. Total fx = 31. By David Healy.

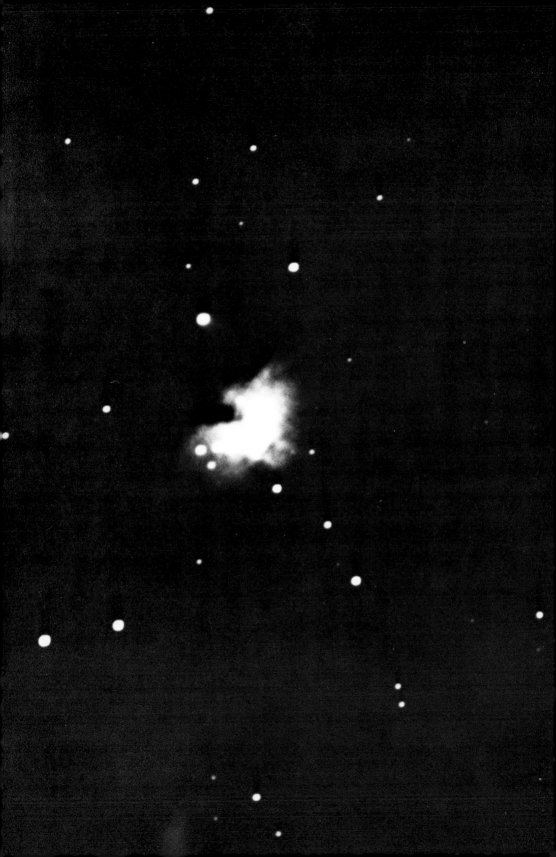

Photograph 11-8. NGC 2244, the Rosette Nebula. Taken November 1978 in Naco, AZ on Kodak 103a·E, 300 mm $f/1.5$ Schmidt Camera, 10 min, using #92 red filter. Assuming ISO 500 for film, and ignoring filter, total fx = 35. By David Healy.

Photograph 11-9. NGC 1499, the California Nebula. Taken October 1977 in Naco, AZ on Kodak 103a·E, piggybacked 135 mm $f/2.5$, 48 min. Assuming ISO 500 for film, total fx = 36. By David Healy.

Photograph 11-10. NGC 6960, western part of the Veil Nebula or Cygnus Loop. Taken July 1978 in North Sandwich, NH on Kodak 103a·E, 1000 mm $f/5$ (via 0.5x focal reducer on 2000 mm $f/10$ Schmidt-Cassegrain telescope), 45 min. Assuming ISO 500 for film, total fx = 34. By David Healy.

Photograph 11-11. NGC 6960-79-92-95, the Veil Nebula or Cygnus Loop. Taken October 1978 in Naco, AZ on Kodak 103a·E, 300 mm $f/1.5$ Schmidt Camera, 13 min. Assuming ISO 500 for film, and ignoring #92 red filter, total fx = 35. By David Healy.

Photograph 11-12. NGC 7000 and IC 5067-70, the North America and Pelican Nebulae. Taken November 1979 in Naco, AZ on Kodak 103a·E, 300 mm $f/1.5$ Schmidt camera, 15 min, using #92 red filter. Assuming ISO 500 for film, and ignoring filter, total fx = 36. By David Healy.

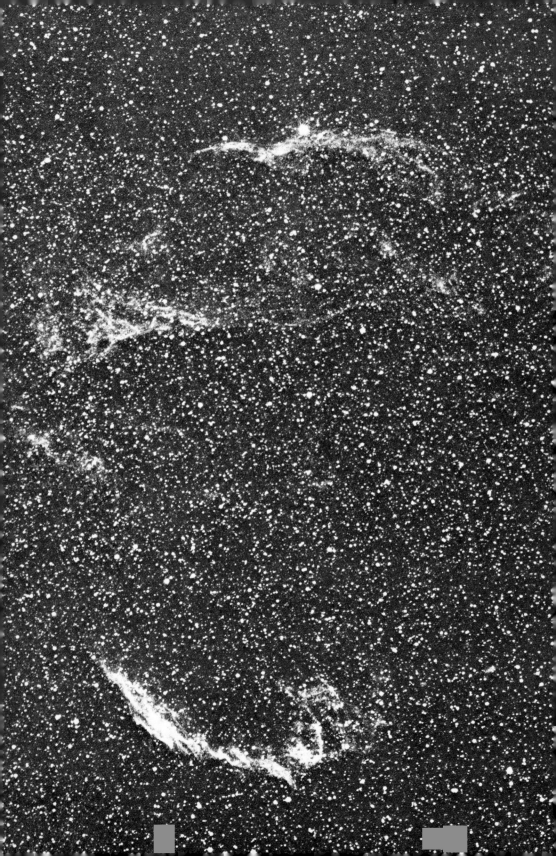

Photograph 11-13. IC434, the Horsehead Nebula. Taken January 1982 in Naco, AZ on gas-hypered Kodak TP2415, 2300 mm $f/6.5$ (via focal reducer on 3900 mm $f/11$ Schmidt-Cassegrain telescope), 80 min. Assuming ISO 250 for film, total fx = 33. By David Healy.

OPEN CLUSTERS

The Messier Catalog lists 27 open star clusters. Their average size is 27' of arc across, but the range is from 6' to a huge 120' — this last for M45, the Pleiades, best-known of the class. As a class, these objects are relatively bright in integrated magnitude, the brightest being M45, followed by M44, the Beehive.

Keep in mind that the open cluster, unique in the Messier Catalog, is not an object to the astrophotographer. Therefore, the magnitude and size data for open clusters really tell us very little about how to photograph them. For example, we find that M45 has a magnitude of about 1.5, meaning a total combined light output equivalent to a star of that magnitude, from a group whose brightest stars are one of third magnitude and five others of fourth magnitude. However, what we are really dealing with here is the group of individual stars, fully resolved, each a point source with its own magnitude and its corresponding individual fx' value. True, the size data will tell us how much sky the group occupies to aid us in focal length selection, but that's really all. We have no fx values; we have no surface brightness; we have no surface. A partial exception to this is the Pleiades, M45. In this case, several brighter stars of the cluster are embedded in nebulosity believed to be leftover fragments of the diffuse nebula from which they were formed. This nebulosity constitutes a worthy challenge for your camera, requiring good dark skies and an exposure of about fx = 33.

But remember: for purposes of photography, open cluster integrated magnitudes tell you essentially nothing. An open cluster is merely a group of individual stars being neighborly.

Photograph 11-14. M11 and star clouds in Scutum. Taken November 1979 in Naco, AZ on Kodak SO115, 300 mm $f/1.5$ Schmidt Camera, 12 min. Assuming ISO 100 for film, total fx = 33. By David Healy.

Photograph 11-15. M37 in Auriga. Taken October 1983 in Naco, AZ on gas hypered Kodak TP2415, 2100 mm $f/6$ (via focal reducer on 3900 mm $f/11$ Schmidt-Cassegrain telescope), 25 min. Assuming ISO 250 for film, total fx = 31. By David Healy.

Photograph 11-16. M45, the Pleiades, "visited by" (a somewhat overexposed) Jupiter. Taken February 1977 in Everglades National Park, FL on Fujichrome F100, piggybacked 85 mm f/1.8, 6 min. Total fx' = 18'.

GLOBULAR CLUSTERS

Globular clusters are also groups of stars, differing from open clusters in several respects. While the open clusters are loose collections of several hundred stars relatively near us, globulars are spherical aggregates of hundreds of thousands of stars vast distances from us. This combination of huge star populations and vast distances makes resolution into individual stars out of the question, except around the edges. Thus, while it is true that globulars are in fact "simply groups of stars," for our purposes a globular cluster is a single object, and an extended object at that.

The Messier Catalog lists 29 globular clusters. This is the most uniform of the three major classes in the Messier Catalog: integrated magnitudes ranging from 6 to 9, and diameters ranging from about 4' to 23' of arc. Data is given in Table 11-3 for the northern and the southern showpiece of the class.

	fx	Mag.	A'
M13 — Hercules	31 to 32	6	23
NGC 5189 — omega Centauri	32	4	65

Table 11-3. *fx* values, integrated magnitudes, and sizes for a northern and southern hemisphere showpiece globular cluster.

Photograph 11-17. M13, the Hercules Cluster. Taken June 1981 in Manhasset, NY on Kodak 103a·E, 1000 mm *f*/5 (via 0.5x focal reducer on 2000 mm *f*/10 Schmidt-Cassegrain telescope), 5 min. Assuming ISO 500 for film, total *fx* = 31. By David Healy.

Photograph 11-18. M45, the Pleiades. Taken November 1974 in Naco, AZ on Kodak SO115, 300 mm *f*/1.5 Schmidt camera, 30 min. Assuming ISO 100 for film, total *fx* = 34.5. By David Healy.

Photograph 11-19. NGC 869-84, the Double Cluster in Perseus. Taken August 1977 in North Sandwich, NH on Kodak 103a·E, piggybacked 300 mm *f*/4, 20 min. Assuming ISO 500 for film, total *fx* = 33. By David Healy.

GALAXIES

The ultimate subject, galaxies are "island universes" consisting of stars and all of the other deep-sky objects. In star population and distance both, they dwarf everything we have considered. No resolution into stars is possible, even at the edges, except for the largest observatory instruments. Thus we again deal with an aggregation as though it were a single extended object. Again we deal with a class of object characterized by a very great brightness variation from center to edge.

With 40 entries this class of object is the largest of the five into which we have divided the Messier Catalog. And a varied lot they are. First, the group includes elliptical galaxies and irregulars, in addition to the showpiece spirals. Second, the integrated magnitudes of the group range from 5 to 11, averaging around 9. Finally, their apparent sizes run from a very modest 2' by 2' to an immense 160' by 40', averaging around 12' by 6' of arc. Table 11-4 is a galaxy sampler, giving *fx* values, plus integrated magnitudes and sizes in arc minutes.

	fx	Mag.	A'
M31 — Andromeda			
Center	31 to 32		
Overall	32 to 33	5	160 by 40
M33 — Triangulum	32 to 33	6	60 by 40
M51 — Whirlpool	33	9	12 by 6
Small Magellanic Cloud	34	3	240 by 216
Large Magellanic Cloud	34	1	480 by 480
"Typical" Galaxy	35	–	–
Milky Way	35	–	–

Table 11-4. *fx* values, integrated magnitudes, and sizes for a selection of galaxies.

Now you know where that hypothetical *fx* = 35 subject discussed earlier in the chapter came from. This subject group demands your longest exposures and darkest skies. But when you've mastered the galaxies, you've mastered it all!

Photograph 11-20. M31, the Great Galaxy in Andromeda. Taken November 1979 in Naco, AZ on Kodak SO115, 300 mm $f/1.5$ Schmidt Camera, 30 min. Assuming ISO 100 for film, total fx = 34.5. By David Healy.

Photograph 11-21. Inner portion of M31, the Great Galaxy in Andromeda. Taken October 1983 in Naco, AZ on gas hypered Kodak TP2415, 2100 mm $f/6$ (via focal reducer on 3900 mm $f/11$ Schmidt-Cassegrain telescope), 40 min. Assuming ISO 250 for film, total fx = 32. By David Healy.

Photograph 11-22. M33 in Triangulum. Taken September 1981 in Naco, AZ on gas hypered Kodak TP2415, 2500 mm $f/7$ (via focal reducer on 3900 mm $f/11$ Schmidt-Cassegrain telescope), 61 min. Assuming ISO 250 for film, total fx = 32. By David Healy.

Photograph 11-23. M51, the Whirlpool Galaxy. Taken May 1981 in Naco, AZ on gas hypered Kodak TP2415, 1950 mm $f/5.5$ (via 0.5x focal reducer on 3900 mm $f/11$ Schmidt-Cassegrain telescope), 35 min. Assuming ISO 250 for film, total fx = 32. By David Healy.

Photograph 11-24. M81 in Ursa Major. Taken February 1981 in Naco, AZ on gas hypered Kodak TP2415, 1950 mm $f/5.5$ (via 0.5x focal reducer on 3900 mm $f/11$ Schmidt-Cassegrain telescope), 40 min. Assuming ISO 250 for film, total fx = 32. By David Healy.

Photograph 11-25. M101 in Ursa Major. Taken March 1977 in Naco AZ on Kodak 103a·E, 1000 mm $f/5$ (via 0.5x focal reducer on 2000 mm $f/10$ Schmidt-Cassegrain telescope), 60 min. Assuming ISO 500 for film, total fx = 34. By David Healy.

Photograph 11-26. NGC 253 in Sculptor. Taken September 1981 in Naco, AZ on gas hypered Kodak TP2415, 2500 mm $f/7$ (via focal reducer on 3900 mm $f/11$ Schmidt-Cassegrain telescope), 40 min. Assuming ISO 250 for film, total fx = 32. By David Healy.

Photograph 11-27. NGC 5128, Centaurus A. Taken April 1984 in Naco, AZ on gas hypered Kodak TP2415, 2100 mm *f*/6 (via focal reducer on 3900 mm *f*/11 Schmidt-Cassegrain telescope), 48 min. Assuming ISO 250 for film, total *fx* = 32. By David Healy.

Photograph 11-28. NGC 4565 in Coma Berenices. Taken April 1984 in Naco, AZ on gas hypered Kodak TP2415, 2100 mm *f*/6 (via focal reducer on 3900 mm *f*/11 Schmidt-Cassegrain telescope), 60 min. Assuming ISO 250 for film, total *fx* = 33. By David Healy.

Chapter 12

EXOTICA

If it has all been mastered, what is there left to discuss? Some rather sophisticated equipment and techniques for doing even better, or perhaps for doing as well with a bit less effort.

THE SCHMIDT CAMERA

For many subjects—star fields, comets, and the larger deep-sky objects—our typical telescopes are still quite slow and have very limited fields. Even focal reducers don't help much. What we'd really like is an optical system giving good coverage over several degrees, with a speed of around $f/2$ or so. True, camera lenses come that fast, but lenses of that speed tend to be too short for our purposes, and they tend to produce miserable star images.

Well then, how about a good telescope objective of about 300 mm focal length with the desired $f/2$ speed? First of all, such a lens would be quite expensive, to put it mildly. In addition, it would be subject to severe aberration problems. In fact, it would have the same problems inherent in the construction of high-speed camera lenses, but greatly aggravated by a 150 mm aperture.

This gets us back to mirrors. Now, a 150 mm diameter mirror is hardly a big deal. It would seem that all we have to do is make our mirror an $f/2$ to automatically get the desired wide field and high speed. That is indeed the case. Unfortunately, most of that wide field is so afflicted with coma as to be absolutely useless. The central portion is fine, as are the central portions of images formed by parabolic mirrors of any speed, coma being an off-axis aberration. How bad is it? Well, again there is no hard and fast answer. As a rule of thumb, the "coma free field" formed by a parabolic mirror, in millimeters, will be somewhat less than the square of the mirror focal ratio. So our common $f/8$ mirrors give fine images over fields approaching 64 mm in diameter, and an $f/6$ mirror will still cover our 35 mm film frame with a coma-free field close to 36 mm in diameter. The 200″ Hale Telescope on Palomar Mountain, with its $f/3.3$ mirror, would give a usable image only a mere 10 mm across, were it not for its sophisticated, and expensive, special correcting lens. An $f/2$ parabolic mirror would offer less than 4 mm of coma-free image!

That being the case, why not go back to spherical mirrors? Since any line perpendicular to a spherical surface is effectively an axis thereof, and since one such ray line will be included in each bundle of parallel rays from any direction, it follows that, to a spherical mirror, all points of the image are "on-axis" points, so that no off-axis aberrations are possible. Spherical mirrors are much easier to make; in fact, that's how all mirrors start out, becoming paraboloids only in the final stages of manufacture. The answer to that last question is simply that spherical mirrors just don't focus light properly, suffering from a flaw called spherical aberration, which explains parabolizing for mirrors. Unfortunately, parabolizing just trades one failing for another, coma which is particularly bad when relatively short focus is desired.

In 1930, a genius named Bernhard Schmidt decided that there should be some way to simply eliminate spherical aberration and not replace it with coma. He did that by introducing a thin, specially shaped corrector plate at a specific distance ahead of his spherical mirror. The results were spectacular: no spherical aberration, no coma, no chromatic aberration, no astigmatism, no image-degrading aberrations at all—at any focal ratio desired, with superb star images across the entire field. In short; pure, unadulterated bliss. Almost.

The first problem is that these superb star images do not lie in a plane. The focal surface of Schmidt's spectacular optics, like his mirror, is a portion of a sphere. That, incidentally, is not unique to this specific optical system. Many lenses tend to have curved focal surfaces; flattening those focal surfaces to conform to a film plane is one problem faced by designers of camera lenses. But, here again, Schmidt's genius asserted itself. Rather than compromise his virtually perfect optics he simply said: leave the lens alone; curve the film. Since the curve of the focal surface was gentle enough, that is what was done. Special film holders produce the proper curve in the film.

The final problem, now that we've accommodated to the focal surface's spherical shape, is its location. Rather than be situated behind the optics where it can easily be reached, the Schmidt camera focal surface is located right smack in the middle—midway between corrector plate and mirror: as much in the way as possible for our precious light rays, and as out of the way as possible for the photographer. Like it or not, that's where it is, and, with the exception of a few sophisticated and complex designs, that's where it seems destined to remain.

Though the early Schmidt cameras were custom-made by skilled amateurs, a number of rather good ones are now available from commercial manufacturers. They tend to run from 225 mm to 300 mm, with speeds in the $f/1.5$ to $f/2.5$ range. Thus, these marvels will cover about 5°

of sky at a time, and, as far as extended objects are concerned, they'll do it in about one-thirtieth the time required by conventional optics. That's another ballpark figure, but the contrast is striking, whatever the precise ratio.

A slight drawback to the Schmidt camera is suggested by its name: that is, it is purely a camera with no means of visual observing. Thus, off-axis guiding is not possible, and a separate guidescope is a must. The worst drawback is the crazy arrangement of special film holders sitting in the middle of the tube. This means the handy 35 mm SLR body is out of the picture.

Schmidt camera users traditionally started with a special film cutter, with which they cut up rolls of 35 mm film into little film "chips" of the proper size, to be carefully placed into those curved film holders—all in total darkness. The worst came after those film "chips" were exposed: no lab was about to tackle the processing of odd bits of film. That meant that processing became strictly your own concern, and, unfortunately, provided considerable opportunity for totally ruining precious exposures. But for those willing and able to deal with the added film-handling chores, Schmidt camera photography produced some truly beautiful results.

Equipment and techniques are now available that offer the option of somewhat simplified Schmidt camera film handling. While the awkward film shape and location remain unchanged, a more elaborate film holder has been developed. This new holder is somewhat larger than the original and more angular in shape, which results in a slight loss of image brightness and resolution. It does, however, permit the use of complete rolls of film, which facilitates processing.

Either way, rolls or "chips", guiding times are trivial by comparison with other astrophoto technology. However, at the modest focal lengths available, this technique applies only to wide-field photography. We must seek answers to our long-focus photo problems elsewhere.

THE TRICOLOR PROCESS

Reciprocity failure does not merely aggravate already long guiding times, it tends to degrade color rendition. This is because the three monochrome films in the color film sandwich do not all suffer reciprocity failure in precisely the same way. A solution to this problem is to make your own color film. That is, take three separate monochrome exposures of your subject, using film and filter combinations that confine each exposure to one primary color of light. In this technique, the films used can be special astronomical (black-and-white) emulsions with very good low-light reciprocity characteristics. Thus, the total time for the three exposures needed will be comparable to that of a single exposure on a typical color film.

The big advantage of this technique is the control it offers in color quality. The big drawback is the considerable lab work involved in producing the final image from three original images, each dyed its appropriate color. While results can be breathtaking, this technique is only for the skilled and dedicated lab worker.

THE COLD CAMERA

Another approach to the reciprocity failure problem came with the discovery that films retain much of their normal sensitivity, even in low-light astronomical use, if chilled to rather low temperatures during exposure. Even normal commercial color films do well at temperatures achievable with solid carbon dioxide, also known as dry ice. Since dry ice is not difficult to obtain, this looked like the solution! no film chips to handle, no three-exposure sets to dye and register, just a container of dry ice to be held against the back of a camera body.

That is all there is to it—if you're photographing from outer space. Trouble is, here on earth our old nemesis the atmosphere clobbers us in a brand new way. As soon as that film starts cooling down, water vapor from the surrounding air begins to condense on its surface. That being intolerable, the cold camera was developed. In one form, it includes a plastic block sitting against the film and occupying the space immediately before it. This prevents condensation until the plastic itself cools enough for its forward surface to become afflicted; meanwhile, it simply tends to degrade image quality. A more effective solution is a vacuum chamber just ahead of the film, which adds the chore of evacuating said chamber, in addition to the loading of the dry ice reservoir, in addition to all the normal tasks of astrophotography.

As with the previous techniques, very fine results are possible, but they don't come easily.

BAKING

It may sound ridiculous, but baking the emulsion is also quite effective in attacking reciprocity failure; but the baking is done prior to, rather than during, exposure. Astronomers use this technique extensively. Another thing they use is glass plates. While the emulsion is improved by the baking process, the glass plate supporting that emulsion is totally unaffected. Plastic backing that support film emulsion is a totally different story as it does not take kindly to baking. When you get serious enough to start exposing glass plates, you might consider baking for improved sensitivity.

SOAKING

The third technique for hypersensitizing—i.e., increasing the effective speed of an emulsion, usually called hypering—is "soaking." Also known as gas hypering, the technique is to soak the film in some gas before exposure. It was found that leaving the film in a pure hydrogen environment for some time before use had a beneficial effect on speed and reciprocity characteristics. Unfortunately, hydrogen has this anti-social tendency to exhibit what is often referred to as the Hindenburg Effect. All in all, it is not a very inviting substance.

However, it has been found that an alternative exists, called forming gas, which contains so little hydrogen as to be quite safe. This has made gas hypering a more attractive technique. Kits are now commercially available which include the gas plus the soaking chamber plus the pump and required fittings. Better still for the uncommitted, it is possible to buy film which has already been gas-hypered.

Since available films and their specifications change so often, and since different films react differently to hypering, it is just not feasible for a book to offer specifics on the subject. As noted earlier, the astronomy periodicals which serve the amateur are a good source of data. Another good data source is the suppliers of hypered film. As a last resort, lacking any data at all, start out by exposing hypered film as though it were $fx = 5$—corresponding to ISO 500.

Here at last is something that seems to offer significant advantages to the astrophotographer at relatively slight additional burden. Of all the hypering techniques, soaking seems to be the one with real value for a fairly sizable number of us.

Thus, to wrap up this brief overview of exotica, gas hypering seems a good next step for the full-fledged astrophotographer. Beyond that, the Schmidt camera and the tricolor process each offer a supply of rewards and challenges that should be sufficient to satisfy even the most avid top-level practitioner.

IN CONCLUSION

We have covered a lot of territory—from the meteors at the outer reaches of our atmosphere, to the galaxies at the outer reaches of the known universe. We have considered a lot of mechanism—from the family camera to the Schmidt camera. Why?

For some of us, there is the hope of a real contribution to astronomy, perhaps a valuable shot of a supernova. For some of us, there is the hope for immortality, perhaps a comet bearing our name. But for all of us, there is literally an entire universe of incredible beauty waiting to be captured and shared.

As we said at the outset, many astrophotographic subjects are extremely easy to capture and many are extremely difficult. The table on page 203 recaps our subjects, with a reminder as to where they were discussed.

As your skills improve, you will receive more praise for your work, particularly from fellow astrophotographers who can really appreciate it: comments like "Nice exposure judgment" or "Rather competent guiding" and the like. But once in a while, from colleague or layperson, you'll get a simple "Wow!" Whether for Saturn at f = 15,000, or Venus visited by a crescent moon at f = 135, or a Schmidt portrait of M31, or whatever, that simple "Wow!" is perhaps the highest praise of all.

May you receive your fair share.

Appendix A
MATHEMATICAL RELATIONSHIPS UNDERLYING THE *fx* SYSTEM

As stated in Chapter 2, the *fx* system was created with three pre-established requirements to be satisfied:

1. A larger *fx* value indicates a greater effective exposure
2. A difference of 1 in *fx* value indicates a difference equivalent to 1 *f*/stop in effective exposure, i.e., a halving or doubling of effective exposure
3. All exposure problems are solved in the system by addition and/or subtraction of simple numbers

As the mathematically knowledgeable will recognize, the second of these three criteria—requiring adding or subtracting 1 in place of multiplying or dividing by 2—strongly suggests that the logarithm of 2 is central to the scheme. And, indeed, so it turns out, at least with film speed and exposure duration. In dealing with lens aperture, relative or absolute, our logarithm base of $\sqrt{2}$ simply recognizes that exposure varies as the square of the aperture diameter. Density, already an exponential/logarithmic function, requires only a change of scale.

The first criterion simply determines the algebraic signs that relate the *fx* values to the physical quantities, and the third suggests the arbitrary constants chosen to fix the scale origins at agreeable locations.

The following, then, are the resulting equations that relate each of the photographic variables to its *fx* values:

$$fx \;=\; 12 \;+\; \log_2 [\text{duration in seconds}]$$

$$\text{duration in seconds} \;=\; 2^{\,fx \,-\, 12}$$

$$fx \;=\; 10 \;-\; \log_{\sqrt{2}} [\text{focal ratio}]$$

$$\text{focal ratio} \;=\; \sqrt{2}^{\;10 \,-\, fx}$$

$$fx = \log_{\sqrt{2}} \text{[aperture in cm]} - 10$$

$$\text{aperture in cm} = \sqrt{2}^{\,fx + 10}$$

$$fx = \log_2 \text{[ISO]} - 4$$

$$\text{ISO} = 2^{\,fx + 4}$$

$$fx = -\frac{10}{3} \text{[density]}$$

$$\text{density} = 0.3\,fx$$

Of course, the above equations simply relate each of the variables under our control to its *fx* value. The missing piece of the system is the relationship between subject brightness and its corresponding *fx* total. This link is an either/or, depending on whether we're dealing with extended objects or point sources. The latter are treated in detail in Chapter 5 and Appendix D, which delve more deeply into considerations of magnitude. The former are handled as follows:

$$B\ I\ t = r^2\ 10^D \quad \text{(for extended object only)}$$

$$fx = 18 - \log_2 B$$

$$B = 2^{18 - fx}$$

Appendix B

DISTANCES ON THE CELESTIAL SPHERE

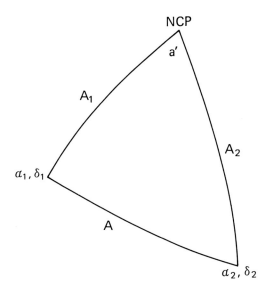

Figure B-a. Distances on the Celestial Sphere.

The distance between the north celestial pole and any point on the celestial sphere — with right ascension and declination of α and δ, respectively — is simply the complement of the point's declination, or 90-δ in degrees. Note that this works for both hemispheres, as southern hemisphere declinations are negative, thus giving distance values greater than 90° where called for.

For any two points other than the NCP (north celestial pole), we can find the distance between them by considering them to be two vertices of a spherical triangle, with the NCP as the third vertex of that triangle. As shown above, A_1 and A_2 represent the distances from the NCP to the points, forming two sides of our spherical triangle. From the definition of right ascension, the angle a' included between these two sides is given by:

$$a' = |\alpha_1 - \alpha_2|$$

which is simply the difference, dropping sign, of the right ascensions of the points. The sign can be dropped as this quantity is used only for finding a cosine — and cos(-x) = cos x for all x. However, right ascension will have to be converted from "time" units (hrs, min, sec) into angular units (degrees) in order to obtain that cosine value.

The quantity we seek, A, is the third side of our spherical triangle. This is given by the Law of Cosines for spherical triangles:

$$\cos A = (\cos A_1 \cos A_2) + (\sin A_1 \sin A_2 \cos a')$$

However, we already know that

$$A_i = 90 - \delta_i$$

Further, from elementary trigonometry, we also know that

$$\cos (90 - \delta) = \sin \delta \quad \text{and} \quad \sin (90 - \delta) = \cos \delta$$

or

$$\cos A_i = \sin \delta_i \quad \text{and} \quad \sin A_i = \cos \delta_i$$

Thus, we can rewrite our formula in its final, most convenient form:

$$\cos A = (\sin \delta_1 \sin \delta_2) + (\cos \delta_1 \cos \delta_2 \cos a')$$

Note that the formula is valid for any point, including the celestial poles which may have any right ascension we wish to assign. However, for the special case

$$\alpha_1 = \alpha_2$$

it is far easier to observe that, since a' = 0:

$$\cos a' = \cos 0 = 1$$

so that

$$\cos A = (\sin \delta_1 \sin \delta_2) + (\cos \delta_1 \cos \delta_2)$$

Thus, from the formula for the cos of the difference of two angles

$$A = \delta_1 - \delta_2$$

which is precisely what we were hoping for.

Appendix C
IMAGE SIZE AND RELATED DATA

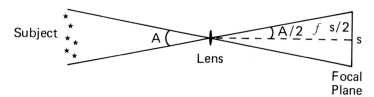

Figure C-1. The Geometry of Image Size.

IMAGE SIZE

Given a subject of apparent angular size A, and a distortion-free lens of focal length f, let s represent the size of the image formed. Since it is a fundamental property of distortion-free lenses that light rays passing through the optical center continue on to the focal plane undeflected, we have

$$\frac{s}{2} = f\left(\tan \frac{A}{2}\right)$$

or

$$s = 2f\left(\tan \frac{A}{2}\right)$$

For small angles, the tangent of the angle approaches the angle itself expressed in radians. Since 1 radian = $(180/\pi)°$ which is approximately $57.3°$ we can replace the tangent function by the factor $1/57.3$. Thus:

$$s \approx f\left(\frac{A}{57.3}\right)$$

173

for A in degrees.

For $A < 20°$ the error in this approximation will be less than or equal to 1%.

FORMAT COVERAGE

Adopting 22.9 mm by 34.2 mm as the image area of a mounted 35 mm slide, we can substitute these values for s and solve for A in both our exact equation and our approximation. This will show our 35 mm format coverage to be:

$$2 \arctan \frac{11.45}{f} \quad \text{by} \quad 2 \arctan \frac{17.1}{f}$$

or approximately:

$$\frac{1312}{f} \quad \text{by} \quad \frac{1960}{f}$$

for f in mm. For f greater than 100, the errors in these values will be less than 1%.

IMAGE TRAILING

Given an exposure duration in seconds of t, and a subject with declination δ, let d, for drift, represent the length of the image trail formed by the subject as a result of using a stationary camera. Since the earth rotates at the rate of 15° per hour, it will rotate $(15/3600)° = (1/240)°$ each second. Thus, the apparent angular size of trail, in degrees, will be:

$$\frac{t}{240} \cos \delta$$

and, from our image size formula:

$$d = 2f \tan \left(\frac{t}{480} \cos \delta \right)$$

Replacing the tangent function by the factor 1/57.3 and multiplying constants, we come out with a final factor of approximately 1/13750. Thus, our approximation for image trailing is:

$$d \approx \frac{f\,t}{13750}\,\cos\,\delta$$

When t is less than 4800 seconds, the error here will be equal to or less than 1%.

If we solve for t in both the exact equation and the approximation, we find:

$$t = \frac{480\,\arctan\,\dfrac{d}{2f}}{\cos\,\delta}$$

and:

$$t \approx \frac{13750\,d}{f\,\cos\,\delta}$$

Where d is equal to or less than $f/3$, the error in this approximation will be less than 1%.

Worst-case trailing occurs at $\delta = 0°$, giving cos $\delta = 1$, i.e., at the celestial equator. Trailing drops to about 3/4 of this at $\delta = 40°$, exactly 1/2 at $\delta = 60°$, and about 1/4 at $\delta = 75°$. Conversely, compared to equatorial trailing, a given trail length would take about 4/3 as long at $\delta = 40°$, exactly twice as long at $\delta = 60°$, and about 4 times as long at $\delta = 75°$.

Appendix D
fx VALUES FOR POINT SOURCES

Since the *fx* scale is based on doubling (i.e., powers of 2) successive *fx* values denote relative effective exposures in the sequence

$$1, 2, 4, 8, 16, 32, 64, \ldots$$

Magnitudes, on the other hand, are based on powers of the fifth root of 100, roughly 2.512, which generates a brightness sequence of approximately

$$1, 2.5, 6.3, 16, 40, 100, 250, \ldots$$

For all practical purposes, a magnitude change of three is equivalent to a change of four in *fx* value, since each represents a change in a ratio very close to 16:1. Thus, *fx* units being somewhat smaller than magnitude units, it takes 4/3 as many of them to offset a given difference in magnitude.

Stated algebraically,

$$3 \; fx \; \approx \; 4 \; mag \; + \; k$$

where k is a constant that must be determined.

Switching to *fx'* notation to emphasize the fact that we're dealing with point sources, we note that experience suggests that stars of magnitude ≈ 1 do very well with exposures of *fx* ≈ 10. Thus, we can say

$$3 \times 10 \; \approx \; 4 \times 1 \; + \; k$$

and solve for k. The result is k ≈ 26.

Substituting back into our first equation gives

$$3 \; fx' \; \approx \; 4 \; mag \; + \; 26$$

giving

$$fx' \approx \frac{4\ \text{mag} + 26}{3}$$

as our approximation for converting magnitudes into fx' values. Since no real precision is possible or necessary here, we shall adopt the slightly simpler

$$fx' \approx \frac{4}{3}\ \text{mag} + 9$$

as our working approximation for this conversion.

For devotees of direct computation, there is an equation equivalent to the above which gives approximate point source exposures without the use of tables. It is the point source analog of the BItrD equation of Chapter 2, with I, t, and D, respectively, representing ISO film speed, seconds of exposure, and attenuator density. However, dealing with point sources, we drop B and r, using instead M for subject magnitude and a for absolute aperture (in centimeters). The complete formula is:

$$I\ t\ a^2 = 100^{\frac{2M + 5D + 16}{10}}$$

As usual, this becomes simpler when D = 0:

$$I\ t\ a^2 = 100^{\frac{M + 8}{5}}$$

Although, as stated, this equation does not really require tables, a short "compromise table" can greatly simplify its use. We simply lump the entire right side into a single (approximate) constant value for each magnitude, as follows:

M	$100^{\frac{M+8}{5}}$
−3	100
−2	250
−1	625
0	1,600
1	4,000
2	10,000
3	25,000
4	62,500
5	160,000
6	400,000
7	1,000,000

The simplified equation now becomes:

$$I\ t\ a^2 \approx \text{constant from table for desired magnitude}$$

This equation yields the same exposures as the *fx* data given above.

Appendix E
LIMITING MAGNITUDE

Let:

$$f \quad = \text{ focal length of lens in millimeters}$$
$$M \quad = \text{ limiting magnitude}$$
$$[\textbf{ISO}] = fx \text{ value of film speed}$$
$$[\textbf{a}] \quad = fx' \text{ value of absolute aperture}$$
$$[\textbf{r}] \quad = fx \text{ value of focal ratio r}$$
$$[\textbf{t}] \quad = fx \text{ value of exposure duration}$$
$$fx \quad = fx \text{ value of sky background}$$
$$fx' \quad = fx' \text{ value of limiting magnitude M}$$

Note that, with f in its customary mm and a in its customary cm, a conversion factor of 10 must appear here and there in our derivation.

If we assume that limiting magnitude stars and sky background are both underexposed by the same amount, u, expressed as a number of $f/$ stops, then

$$[\textbf{t}] + [\textbf{r}] + [\textbf{ISO}] = fx - u$$

and also

$$[\textbf{t}] + [\textbf{a}] + [\textbf{ISO}] = fx' - u$$

Since we are considering a single exposure, $[\textbf{t}]$ and $[\textbf{ISO}]$ are each fixed values for the two equations. Thus, by subtracting the second equation from the first, we get

$$[\textbf{r}] - [\textbf{a}] = fx - fx'$$

From Appendix A, we have

$$[\textbf{r}] = 10 - \log_{\sqrt{2}} r$$

181

and

$$[a] \ = \ \log_{\sqrt{2}} a \ - \ 10$$

Substitution yields

$$(10 \ - \ \log_{\sqrt{2}} r) \ - \ (\log_{\sqrt{2}} a \ -10) \ = \ fx \ - \ fx'$$

which simplifies to

$$fx' \ = \ fx \ + \ \log_{\sqrt{2}} \frac{f}{10} \ - \ 20$$

Converting that log to base ten, and simplifying

$$fx' \ = \ fx \ + \ \frac{2}{\log 2} \ \log f - \left(20 \ + \ \frac{2}{\log 2}\right)$$

If we want this directly in terms of magnitude, we refer to Appendix D to find

$$fx' \ \approx \ \frac{4}{3} M \ + \ 9$$

so that

$$\frac{4}{3} M \ + \ 9 \ \approx \ fx \ + \ \frac{2}{\log 2} \ \log f - \left(20 \ + \ \frac{2}{\log 2}\right)$$

which reduces to

$$M \ \approx \ \frac{3}{4} fx \ + \ \frac{3}{2\log 2} \ \log f - \left(\frac{87}{4} \ + \ \frac{3}{2\log 2}\right)$$

Evaluating the constants yields

$$M \ \approx \ \frac{3}{4} fx \ + \ 5 \ \log f - 27$$

as the approximate (theoretical) limiting magnitude achievable with a given focal length against a sky with given fx value.

Appendix F

BALANCING POINT SOURCE AND EXTENDED OBJECT

Let:

$$
\begin{aligned}
f &= \text{focal length of lens in millimeters} \\
M &= \text{magnitude of point source subject} \\
[\text{ISO}] &= fx \text{ value of film speed} \\
[a] &= fx' \text{ value of absolute aperture} \\
[r] &= fx \text{ value of focal ratio r} \\
[t] &= fx \text{ value of exposure duration} \\
fx &= fx \text{ value of extended object subject} \\
fx' &= fx' \text{ value of point source subject of magnitude M}
\end{aligned}
$$

Note that, with f in its customary mm and a in its customary cm, a conversion factor of 10 must appear here and there in our derivation.

If we assume that point source and extended object are both optimally exposed by the same exposure, then

$$[t] \; + \; [r] \; + \; [\text{ISO}] \; = \; fx$$

and also

$$[t] \; + \; [a] \; + \; [\text{ISO}] \; = \; fx'$$

Since we are considering a single exposure, t and ISO are each fixed values for the two equations. Thus, by subtracting the second equation from the first, we get

$$[r] \; - \; [a] \; = \; fx \; - \; fx'$$

From Appendix A, we have

$$[r] \; = \; 10 \; - \; \log_{\sqrt{2}} r$$

and

$$[\mathbf{a}] \;=\; \log_{\sqrt{2}} a - 10$$

Substitution yields

$$(10 - \log_{\sqrt{2}} r) - (\log_{\sqrt{2}} a - 10) \;=\; fx - fx'$$

which simplifies to

$$\log_{\sqrt{2}} \frac{f}{10} \;=\; fx' - fx + 20$$

And, finally

$$f \;=\; 10 \times 2^{\frac{fx' - fx + 20}{2}}$$

If we want this directly in terms of magnitude, we refer to Appendix D to find

$$fx' \;\approx\; \frac{4}{3} M + 9$$

so that

$$f \;\approx\; 10 \times 2^{\frac{4/3\, M + 9 - fx + 20}{2}}$$

and finally,

$$f \;\approx\; 10 \times 2^{\frac{4M - 3fx + 87}{6}}$$

This, then, is approximately the focal length that (theoretically) will produce equal exposure for a point source with given magnitude and an extended object with given fx.

Appendix G
AUXILIARY LENS MAGNIFICATION

Let us add two more symbols to Figure 8-1, creating Figure G-1. The first is b, which is simply the diameter of the auxiliary lens, corresponding to the aperture a of the prime lens. Obviously, b should appear associated with the auxiliary lenses in each of the lower three diagrams. The second symbol, or pair of symbols, is V′ and V. This symbol designates the angle formed by the (outermost) converging light rays where they meet: V′ for the original focal point; and V for the final one. Thus, V′ should appear in all four diagrams at a distance f ′behind the prime lens, and V should appear in the lower three diagrams at a distance y behind the auxiliary. Note that on the lower three diagrams, V′ is also a distance x from the auxiliary lens, in the bottom diagram, it actually designates the two (equal) angles on either side of the prime focus point.

As noted several times previously,

$$r' = \frac{f'}{a}$$

Therefore

$$r' = \frac{1}{2}\left(\frac{f'}{a/2}\right)$$

$$= \frac{1}{2}\cot\frac{V'}{2}$$

where "cot" is the cotangent function of the angle.

This last equation is actually a perfectly valid, and often quite handy, alternative way to define focal ratio. It has the very real advantage of ignoring those "remote" attributes, f and a, of some lens or lenses off in space somewhere. Instead, it considers only the final rate of convergence of the cone of light, right where the final image is formed, totally disregarding any and all modifications and/or transformations that might have

185

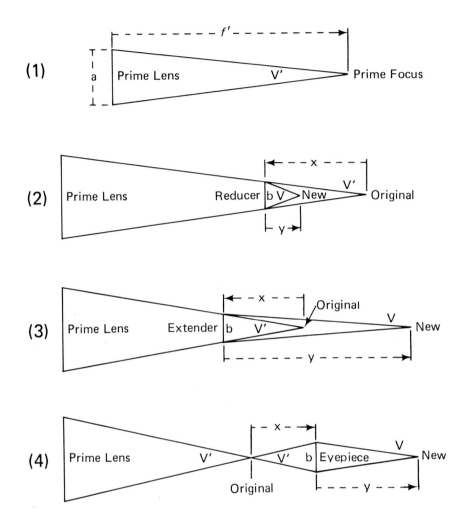

Figure G-1. Optical Systems.

occurred previously. That is just fine, because it is precisely that final convergence angle, and that alone, which defines the final effective focal ratio. Therefore, we can also state

$$r = \frac{1}{2} \cot \frac{V}{2}$$

From these last two

$$\frac{r}{r'} = \frac{\frac{1}{2}\cot\frac{V}{2}}{\frac{1}{2}\cot\frac{V'}{2}} = \frac{\cot\frac{V}{2}}{\cot\frac{V'}{2}}$$

At this point, following the above suggestion, we can ignore prime lens characteristics and focus instead on what our auxiliary lens is doing. Since

$$\cot\frac{V}{2} = 2\,r \quad \text{and} \quad \cot\frac{V'}{2} = 2\,r'$$

this leads directly to

$$\frac{r}{r'} = \frac{\frac{y}{b/2}}{\frac{x}{b/2}} = \frac{y}{x}$$

where x and y are both positive, although, in many other instances, they must be treated as signed quantities according to some acceptable sign convention.

Since

$$f' = a\,r' \quad \text{and} \quad f = a\,r$$

then

$$\frac{f}{f'} = \frac{a\,r}{a\,r'} = \frac{r}{r'} = \frac{y}{x}$$

Recalling our discussion of image size, Chapter 3, we can further state

$$s' = 2f'\tan\frac{A}{2}$$

and

$$s = 2f\tan\frac{A}{2}$$

so that

$$\frac{s}{s'} = \frac{2 f \tan \dfrac{A}{2}}{2 f \tan \dfrac{A}{2}} = \frac{f}{f'} = \frac{r}{r'} = \frac{y}{x}$$

Finally, since (by definition)

$$m = \frac{s}{s'}$$

we have

$$m = \frac{s}{s'} = \frac{f}{f'} = \frac{r}{r'} = \frac{y}{x}$$

Our final way to determine m is in terms of auxiliary lens focal length. It comes directly from a general equation governing all lenses;

$$\frac{1}{\text{focal length}} = \frac{1}{\text{object distance}} + \frac{1}{\text{image distance}}$$

This relationship, a basic one from the field of optics, cannot be applied indiscriminately. For one thing, it is one instance in which we must recognize that our variables, both distances and focal length, are all signed quantities.

Another problem arises when we try to define the distances; strictly speaking, each must be measured from its proper "principal plane", which can be difficult to locate. This is especially true for the compound systems we are creating. However, for "simple" lenses, and relatively simple ones, like our auxiliary lenses, we can assume that both principal planes are essentially at the physical center of the lens. Thus, in the situation we're dealing with, we have

$$\frac{1}{e} = \frac{1}{x} + \frac{1}{y}$$

which is the basic rule relating e, x, and y. This relationship becomes particularly useful when using Eyepiece Projection, a situation in which e is known precisely, and all three variables are positive. It can readily be transformed to

$$\frac{1}{x} = \frac{1}{e} - \frac{1}{y}$$

and then

$$\frac{y}{x} = y \left(\frac{1}{e} - \frac{1}{y} \right)$$

giving

$$\frac{y}{x} = \frac{y}{e} - 1$$

where y is the distance from the eyepiece to the new image plane, and e is the eyepiece focal length. Thus, finally

$$m = \frac{s}{s'} = \frac{f}{f'} = \frac{r}{r'} = \frac{y}{x} = \frac{y}{e} - 1$$

which are the relationships defining the behavior of auxiliary lenses.

Appendix H
GUIDING TOLERANCE

In order to determine guiding tolerances, we will need to work with (among other things) two different focal lengths. We will use the following nomenclature.

T = Guiding Tolerance, as a multiple of box size, i.e., in box-size units
E = Tolerable Pointing/Tracking Error, in some angular measure
B = Apparent Guiding Box Size on Sky, in the same angular measure as E
d = Acceptable Image Drift on Film, in mm
f = Photographic Focal Length, in mm
b = Actual Guiding Box Size, in mm
g = Guiding Focal Length, in mm

We then have, essentially by definition

$$T = \frac{E}{B}$$

Since, from the image size formula of Chapter 3,

$$E = 57.3 \frac{d}{f}$$

and

$$B = 57.3 \frac{b}{g}$$

we have

$$T \approx \frac{57.3 \frac{d}{f}}{57.3 \frac{b}{g}} \approx \frac{d\,g}{b\,f}$$

Note: In using the formula for image size to substitute for E and B, the approximation is more than adequate for our purposes here, since the angles we're dealing with will all be on the order of one arc minute or so.

Answers to Exercises

CHAPTER 2

1. Using *fx*, we need a 12 total, and we get 7 + 2 from relative aperture and film, respectively. The required 3 additional will be provided by a duration of 1/500 second.

2. Focal ratio and film, respectively, provide 4 + 2 toward our required *fx* total of 8. The remaining 2 will come from an exposure duration of 1/1000 second.

3. By the same method as in Exercise 2 above, the answer is 1/30 second.

4. Duration and relative aperture, respectively, give us 12 + 4 toward our required 23 total. The remaining 7 will come from ISO 2000 film.

   Note: In all four of the problems above, lens focal length is absolutely superfluous data.

5. Watch out: point sources are different! Catch: we need absolute aperture in centimeters, and it is not given. So, a 135 mm *f*/2.8 has an aperture of about $135/2.8 \approx 48$ mm — which is 4.8 cm — contributing an *fx* of -5 to -6. Adding the +4 from our ISO 200 still leaves us at -1 to -2. Thus, to reach our required 13, we must add a duration contribution of 14 or 15, which we can get from an exposure of 4 to 8 seconds.

   Moral: The typical camera lens is pretty small-bore stuff for stalking point source quarry.

6. Here, we must arrive at a total of 13 starting with a contribution of 11 + 6 from duration and film, respectively. Thus, the contribution from our lens, absolute aperture, must be -4. We find that such a contribution would come from a lens of 8 cm aperture, but that wasn't what we were asked. To find the focal ratio, we divide focal length by absolute aperture, in the same units. Thus, we have 150/8 in cm or 1500/80 in mm, either one giving a focal ratio of about 19.

Though point sources are not quite the snap that extended objects are, the brighter side of the picture is we really don't have to wrestle with them all that often. When we do want to deal with them, this system gives us a powerful way of doing so. Just remember, for point sources use absolute aperture in cm, not relative aperture in $f/$.

CHAPTER 3

1. Simple substitution in our image size approximation would give:

$$s \approx \frac{50 \times 26}{57.3} \approx 22.7$$

However, since A exceeds $20°$ we might wish to use the exact formula. This substitution gives:

$$s = 2 \times 50 \tan \frac{26}{2} \approx 23.1$$

In this case, the approximate value is low by a bit more than 1.7%, as we are outside the range where 1% precision can be depended on. Whether that is acceptable is a question of how much precision you need and how comfortable you are with tangents of angles.

2. Here we can readily use:

$$s \approx 135 \times \frac{2.7}{57.3} \approx 6.36$$

which we know is very close, certainly within 1%. Actually, it is low by less than 0.04%

3. Careful! The image size given is in arc minutes, which we must convert to degrees by dividing by 60. Down in this range, under $1°$, we know that the approximation is extremely good, so:

$$5 \approx f\frac{(31/60)}{57.3}$$

whence $f \approx 555$ mm

4. Similar to #3 but now dividing by 3600 to convert seconds into degrees:

$$3 \approx f \frac{(48/3600)}{57.3}$$

giving $f \approx 12,900$ mm!

Moral: To turn a planet into an extended object, one must go to great lengths.

BIBLIOGRAPHY

A selection of some of the better
materials published through 1983.

Alt, Eckhard, et al (erroneously attributed only to Kurt Rihm), "Color Portraits of Deep Sky Objects", *Sky and Telescope* (Aug 74) 120–124. Tricolor photography. See also —

Alt, Eckhard, et al, "More About Indirect Color Astrophotography", *Sky and Telescope* (Nov 74) 333–338.

Baumgardt, Jim, "Simple and Inexpensive Tricolor", *Astronomy* (Nov 83), 51–54.

Berry, Richard, "Cold Cameras", *Astronomy* (Feb 77), 50–54, (Mar 77) 42–47. Well-illustrated, two-part article.

Berry, Richard, "Astrophotography with Newtonian Reflectors", *Astronomy* (Sep 77), 46–54. Includes a small astrophoto bibliography to (then) recent issues of *Astronomy*. Also features an excellent discussion of alignment by Robert Provin and Brad Wallis.

Berry, Richard, "Photograph the Moon!", *Astronomy* (Feb 78), 34–39.

Berry, Richard, "Astrophotography — with Camera Only!", *Astronomy* (Jun 78), 42–47.

Berry, Richard, "Exposure in Photography", *Astronomy* (Aug 78), 49–51. The algebraic — or BItrD — approach to exposure. Also includes a graph for determining (approximate) focal lengths for various subject and image sizes.

Berry, Richard, "Piggybacking — Without a Drive", *Astronomy* (Apr 79), 32–35. A simple manual drive for equatorial mountings without one.

Brown, G P, et al, "An Evaluation of Films for Astrophotography", *Sky and Telescope* (May 80), 433–439. Excellent article by three gentlemen from Eastman Kodak, including well known authority George Keene.

Brunk, Berry, "Focus on Luna", *Astronomy* (Sep 82), 50–54. Brief but reasonably complete coverage of subject #1.

Brunk, Berry, and Burnham, Robert, "How Much Sky Can My Camera Capture?", *Astronomy* (Jun 82), 52–54. Introduction to image size and format coverage. (Note, however, that only approximate formulas are used, so that values given for shorter lenses are fairly wide of the mark.)

Burnham, Robert, "Getting the Correct Exposure", *Astronomy* (Jun 81), 51–54. Another look at the algebraic — or BItrD — approach to exposure. (Note that the article's second and third illustrations have inadvertently been switched.)

Burnham, Robert, "Easy Steps to Perfect Polar Alignment", *Astronomy* (Aug 82), 52–54.

Burnham, Robert, "Photographing the Giant Planets", *Astronomy* (May 83), 35–38. . . . i.e., Jupiter and Saturn.

Chou, B Ralph, "Safe Solar Filters", *Sky and Telescope* (Aug 81), 119–121.

Coombs, Lee C, "Get the Most from Your Deep Sky Negatives", *Astronomy* (Oct 80), 50–53. Copying and printing techniques.

Custer, Clarence P, "Photographic Polar Alignment of an Equatorial Mounting", *Sky and Telescope* (Nov 73), 329–333. A technique for precise alignment of a permanently mounted scope.

di Cicco, Dennis, "Filters to Pierce the Nighttime Veil", *Sky and Telescope* (Mar 79), 231–236. 'Light pollution filters.'

di Cicco, Dennis, "Exposing the Analemma", *Sky and Telescope* (Jun 79), 536–540.Perhaps the all-time great story of astrophoto planning — and perseverance — resulting in one (or 48, depending upon how you count) of the truly outstanding astrophotographs.

di Cicco, Dennis, "Notes on Gas Hypersensitizing", *Sky and Telescope* (Feb 81), 176–177. Includes work of Jack Marling.

di Cicco, Dennis, "ASA 1,000 and Color Too!", *Sky and Telescope* (Mar 83), 215–217. Kodacolor 1000, as an astro-film.

di Cicco, Dennis, "Another Superspeed Color Film", *Sky and Telescope* (Dec 83), 506–510. 3M Color Slide 1000, as an astro-film.

Eicher, David J, "Photographing Detail in Galaxies", *Astronomy* (Dec 83), 51–54.

Everhart, Edgar, "Adventures in Fine Grain Astrophotography", *Sky and Telescope* (Feb 81), 100–103. Combining sophisticated techniques and equipment for high quality results.

Felbab, James, "Photographing the Sun", *Astronomy* (Mar 80), 39–43.

Fowler, Thomas B, Jr, "A Nomogram for Astrophotographers", *Sky and Telescope* (May 76), 353–354. A graphical scheme for estimating limiting magnitude.

Gephart, Roy E, "Catch a Falling Star", *Astronomy* (Aug 79), 49–52. Meteor photography.

Haig, G Y, "A Simple Camera Mounting for Short Exposures", *Sky and Telescope* (Apr 75), 263–266. A crude but easily made equatorial mounting that can be hand driven for limited tracking.

Hamler, Walter E, "Equipment for Guiding Your Astrophotos", *Astronomy* (Nov 80), 62-65. Another vote for the 'guiding box' reticle, in a good overview of guiding.

Healy, David, "Astrophotography . . . in Spite of Myself", *Astronomy* (Apr 76), 34–42. A combination of technical data and words of wisdom, self illustrated.

Healy, David, "Astrophotography: Planning Pays Off", *Sky and Telescope* (Feb 79), 197–202. Insights into his work, self illustrated.

Healy, David, "Wide Field Sky Photography", *Astronomy* (Jan 80), 46–50.

Healy, David, "Deep Sky Photography: Films and Exposure Times", *Deep Sky Monthly* (Sep 80), 3–6. Self illustrated.

Healy, David, "Experiments with Gas Hypered Film", *Sky and Telescope* (Feb 81), 174–175.

Healy, David, "Murphy's Astrophotographic Glossary", *Star & Sky* (Mar 79) 18–19. Suggesting that a sense of humor is probably an essential part of an astrophotographer's 'equipment'.

Henzl, Leo C, "Equipment for Guided Astrophotography", *Astronomy* (Jul 75), 50–55.

Henzl, Leo C, "Adequate Astrographs", *Star & Sky* (Sep 80), 70–72. In which proper attention is given to the mounting.

Huffman, Art, "Keeping Warm While Stargazing", *Astronomy* (Dec 80), 36–38. An astronomy teacher deals with a vital-but-oft-ignored subject.

Huffman, Art, "How to Reduce Tracking Error", *Astronomy* (Feb 82), 56–57. A brief introduction to 'the guiding box'.

Iburg, Bill, "The Wet Side of Color Astrophotography", *Astronomy* (Dec 78), 42–45. Printing color astrophotographs.

Jandorf, Hal R, "Photograph the Milky Way", *Astronomy* (Jun 76) 34–40. Illustrated by Mr Jandorf and others.

Jandorf, Hal R, "Short Focus Photography", *Astronomy* (Sep 83), 50–54. Virtues of an RFT (i.e., 'rich field telescope') with focal length of about 500 mm.

Jeffrey, Neil J, "Mathematics in Photography" *Mathematics Teacher* (Dec 80), 657–662. A teacher enlists photography as an aid in teaching math.

Keene, George T, *Star Gazing with Telescope and Camera*, 2nd ed, Amphoto, NY, 1967. An early work in the field, this book deals more with equipment and observing, rather than actual photography. 'Sky-shooting' is the last chapter, 16 pages, with less than one page devoted to guiding.

Lightfoot, Dale, "Making the Most of Black-and-White Astronegatives", *Astronomy* (Jan 82), 51–55. Advanced copying and printing techniques.

Lynch, David K, "Photographing Planetary Surfaces", *Sky and Telescope* (Feb 73), 127–128. Reducing planetary exposures to simple algebra.

Macfarlane, Alan W, "The Astrophotographer's Desiderata", *Sky and Telescope* (Jun 83), 560. Humorous set of astrophoto 'commandments'.

Malin, David F, "Exploring the Image", *Astronomy* (Nov 79), 16–23. Image enhancement via copying or masking.

Malin, David F, "The Colors of Deep Space" *Astronomy* (Mar 80), 6–13. Tricolor technique.

Malin, David F, "Improved Techniques for Astrophotography", *Sky and Telescope* (Jul 81), 4–7. Unsharp masking and 'photographic amplification'.

Malin, David F, "The Deep Sky in Color", *Sky and Telescope* (Sep 81), 216–219. More on tricolor technique.

Marling, Jack B, "The 2415 Revolution", *Astronomy* (Mar 82), 59–63. Good data on an important b&w astrophoto emulsion.

Mayer, Ben, "Improve Your Slides Through Rephotography", *Astronomy* (Feb 79), 34–39. . . . also known as duplicating.

Mayer, Ben, "Starlight and Patience", *Sky and Telescope* (Apr 80), 348. Words of wisdom, rather than technical data.

Melka, James T, and Melvin, Richard, "Suggestions for a Photographic Patrol of Mars", *Sky and Telescope* (Dec 75), 424–428. Very good material on a really challenging subject.

Newton, Jack, "A Cold Camera for Astrophotography", *Astronomy* (Feb 81), 39–42. Very good data, well illustrated.

Newton, Jack, and Hankin, Dale Roy, *Astrophotography — from Film to Infinity*, Astronomical Endeavors, 1974. Brief overview of the entire field, including processing. Also includes notes on solar system photography by John Sanford. Illustrated by Newton, Sanford, and other experts.

Paul, Henry E, *Outer Space Photography for the Amateur*, 4th ed, Amphoto, NY, 1976. Probably the classic work in the field, going back to 1960, with some of the material somewhat dated.

Pike, Robert, "Schmidt Cameras", *Astronomy* (Nov 76), 50–53. Very nicely illustrated discussion of Schmidt optics.

Price, Ronald S, "Planetary Photography at High Resolution", *Sky and Telescope* (Sep 76), 220–223.

Provin, Robert W, and Wallis, Brad D, "Integration Printing: An Aid to Long Exposure Astrophotography", *Astronomy* (Oct 73) 35–39.

Provin, Robert W, and Wallis, Brad D, "Highlight Masking: A Method of Detail Enhancement", *Astronomy* (Dec 73) 19–23.

Provin, Robert W, and Wallis, Brad D, "On the Road to Better Astronomical Photographs", *Sky and Telescope* (Apr 77), 314–318, (May 77), 399–405, (Jun 77), 484–491. A three-part article on the quest for quality by two astrophotographers who really understand the word. Strikingly illustrated by the authors. A gem.

Provin, Robert W, and Wallis, Brad D, "Guiding", *Astronomy* (Dec 80) 39–42. Extremely good information on this all-important subject.

Reed, Steve, "High Resolution Astrophotography", *Astronomy* (Nov 79), 43–48.

Reed, Steve, "Off-axis Guiding: Pain or Panacea?", *Astronomy* (Feb 83), 75–77.

Rouse, James K, "Tell the Planets to Say 'Cheese' ", *Astronomy* (Sep 74), 50–55. Long focus photography.

Rouse, James K, "Essentials of Lunar and Planetary Photography", *Star & Sky* (Feb 80), 20–24, (Apr 80), 36–39, (May 80), 46–49. Three-part article offering a wealth of data.

Royer, Ronald E, "Tricolor Astrophotography," *Astronomy* (Dec 79), 66–71.

Sanford, John, "Moon and Sun Photography — Easy and Satisfying", *Astronomy* (Feb 74), 20–27. Self illustrated.

Sanford, John, "Stationary Camera Astrophotography", *Star & Sky* (Apr 79), 52–53.

Sanford, John, "The Schmidtification of the Universe", *Star & Sky* (Aug 79), 54–55. A mix of history and technical data.

Sanford, John, "Atrophotographic Parameters", *Star & Sky* (Sep 79), 54–55. Some astrophoto mathematics.

Sanford, John, "Guiding", *Star & Sky* (Dec 79), 58–59.

Sanford, John, "Capturing the Planets on Film", *Star & Sky* (Jan 80), 60–61.

Sanford, John, "Boosting Film Speed", *Star & Sky* (Dec 80), 70–72. Gas hypering.

Schwartzenburg, Dewey, "Brightness and Magnitude", *Astronomy* (Jul 79), 54–57.

Sinnott, Roger W, "Further Notes on Exposure Times", *Sky and Telescope* (May 76), 355–356.

Sliva, Roger, "Hypersensitizing, Part 1", *Astronomy* (Apr 81), 39–42. How-to's and comparative results. Part 2, in May 81, tells how to construct your own hypering hardware.

Sliva, Roger, "Getting the Most from Slide Films", *Astronomy* (Jan 83), 35–38. Developing reversal films into negatives.

Smith, Alex G, "New Trends in Celestial Photography", *Sky and Telescope* (Jan 77), 23–28. Hypering techniques.

Strittmatter, Donald J, "A Sunset Silhouette", *Sky and Telescope* (Nov 74). A letter detailing how he obtained the accompanying photograph of Kitt Peak silhouetted against a spectacular, enormous setting sun. A great example of astrophoto planning — and effort — for a striking photograph.

Vehrenberg, Hans, "Photographs of Deep Sky Objects", *Sky and Telescope* (Apr 78) 295–298. Showing what the combination of Schmidt camera and tricolor techniques can do in the right hands.

SUBJECT/CHAPTER CROSS REFERENCE

	Chap 5 Easy Subjects	Chap 7 Driven Camera	Chap 9 Long Lenses	Chap 11 Guided Camera
Star Trails	✔			
Meteor Trails	✔	✔		
Moon	✔		✔	
Star Fields	✔	✔		
Planets	✔		✔	
Conjunctions and Occultations	✔	✔		
Lunar Eclipses	✔	✔		
Sun	✔		✔	
Solar Eclipses	✔			
Transits	✔		✔	
Comets	✔	✔		✔
Planetary Nebulae				✔
Diffuse Nebulae				✔
Open Clusters				✔
Globular Clusters				✔
Galaxies				✔

INDEX

NOTES

NOTES